PROCEEDINGS OF SYMPOSIA
IN APPLIED MATHEMATICS
Volume XXI

MATHEMATICAL ASPECTS OF PRODUCTION AND DISTRIBUTION OF ENERGY

AMERICAN MATHEMATICAL SOCIETY
PROVIDENCE, RHODE ISLAND
1977

PROCEEDINGS OF THE SYMPOSIUM IN APPLIED MATHEMATICS
OF THE AMERICAN MATHEMATICAL SOCIETY

HELD IN SAN ANTONIO, TEXAS
JANUARY 20–21, 1976

EDITED BY

PETER D. LAX

Prepared by the American Mathematical Society
with partial support from NSF grant no. MCS 76-10597
and ERDA contract no. E (11–1)-2409.

Library of Congress Cataloging in Publication Data

Symposium in Applied Mathematics, San Antonio, 1976.
Mathematical aspects of production and distribution of energy.

(Proceedings of symposia in applied mathematics ; v. 21)
Bibliography: p.
1. Power resources--Mathematical models--Congresses. 2. Nuclear engineering--Mathematical models--Congresses. I. Lax, Peter D. II. American Mathematical Society. III. Title. IV. Series.
TJ163.2.S95 1976 333.7 77-7174
ISBN 0-8218-0121-X

AMS (MOS) subject classifications
35Qxx, 62–XX, 65Mxx, 65Nxx, 90Axx, 90Bxx.

CONTENTS

ERRATA

Proceedings of Symposia in Applied Mathematics

Volume 21

"Magnetic confinement fusion energy research"

By Harold Grad

On page 15, Figures 3(b) and 3(c), should be interchanged.

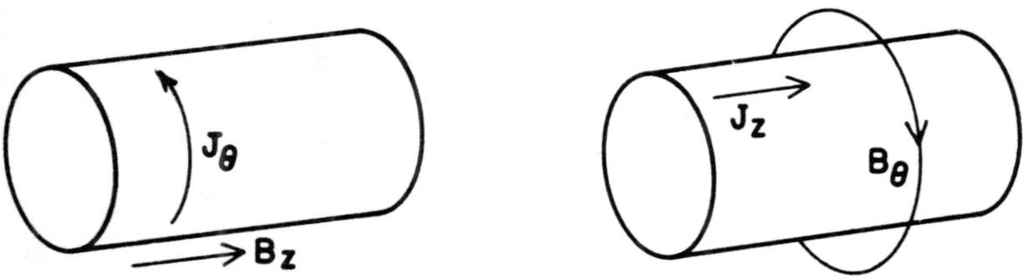

(b) θ-Pinch

(c) Z-Pinch

Figure 3: Cylindrical Equilibria

PREFACE

Volume 21 of the Proceedings of Symposia in Applied Mathematics contains the papers presented at the Energy Short Course held on January 20-21, 1976 in San Antonio, Texas at the Eighty-second Annual Meeting of the American Mathematical Society. The papers are grouped in two categories: those having to do with the mathematical problems involved in the technology of energy production, and those which have to do with the mathematical problems of estimating the resources of energy and the efficient distribution of available energy. In both areas we are dealing with idealized models. The models for energy production are in the form of fairly complicated systems of partial differential equations whose solutions require techniques of finite difference schemes, finite element methods, and Fourier techniques. The models of energy distribution are large networks; their analysis is based on techniques from statistics, linear programming, dynamic programming, and techniques of optimization.

There are more subjects bearing on energy production and distribution than can be fitted into a two-day program; the committee endeavored to select those areas which have most to gain from a mathematical treatment. Some subjects, for instance neutron transport and chemical reactors, were omitted in spite of their great importance for the design of nuclear and chemical reactors because they are already represented in the mathematical literature (see "Transport Theory", SIAM-AMS Proceedings, Vol. I, edited by I.K. Shumays, R. Bellman and G. Birkhoff, American Mathematical Society, Providence, RI, 1969 MR 40 #7062; and "Mathematical Aspects of Chemical and Biochemical Problems and Quantum Chemistry", SIAM-AMS Proceedings, Vol. VIII, edited by D.S. Cohen, American Mathematical Society, Providence, RI, 1974). Other subjects such as solar energy were not represented because the outstanding problems at the present time are technological rather than mathematical.

The organizing committee had asked as speakers mathematicians deeply committed to energy related applications as well as experts in these fields who have a flair for mathematical ideas and techniques. Several of the speakers are not only technical experts in their field, but have also been involved in decision making. For this reason, several of the articles contained interesting remarks of public policy.

The program was under the direction of an Organizing Committee whose members were: George B. Dantzig, Departments of Operations Research and Computer Science, Stanford University; Herbert B. Keller, Firestone Laboratory, California Institute of Technology; Peter D. Lax (chairman), Courant Institute of Mathematical Sciences, New York University; and Robert Marr, Brookhaven National Laboratory. The Short Course was one of a series given by the Society on the recommendation of its Committee on Employment and Educational Policy, consisting of Michael Artin, Charles W. Curtis, Wendell H. Fleming (chairman), Calvin C. Moore, Martha K. Smith and Daniel H. Wagner.

February 1977

Peter D. Lax
Editor

THE MATHEMATICS OF ENERGY PRODUCTION

Proceedings of Symposia in Applied Mathematics
Volume 21
1977

MAGNETIC CONFINEMENT FUSION ENERGY RESEARCH

Harold Grad[1]

ABSTRACT. Controlled Thermonuclear Fusion offers probably the only relatively clean energy solution with completely inexhaustible fuel and unlimited power capacity. The scientific and technological problem consists in magnetically confining a hot, dense plasma (pressure several to hundreds of atmospheres, temperature 10^8 degrees or more) for an appreciable fraction of a second. The scientific and mathematical problem is to describe the behavior, such as confinement, stability, flow, compression, heating, energy transfer and diffusion of this medium in the presence of electromagnetic fields just as we now can for air or steam. Some of the extant theory consists of applications, routine or ingenious, of known mathematical structures in the theory of differential equations and in traditional analysis. Other applications of known mathematical structures offer surprises and new insights: the coordination between sub-supersonic and elliptic-hyperbolic is fractured; supersonic propagation goes upstream; etc. Other completely nonstandard mathematical structures with significant theory are being rapidly uncovered (and somewhat less rapidly understood) such as non-elliptic variational equations and new types of weak solutions. It is these new mathematical structures which one should expect to supply the foundation for the next generation's pure mathematics, if history is a guide. Despite the substantial effort over a period of some twenty years, there are still basic and important scientific and mathematical discoveries to be made, lying just beneath the surface.

INTRODUCTION. In this "Short Course on Energy", this lecture could be considered an even shorter course on "Mathematical Problems Arising in Fusion Energy". In July of 1974 the Courant Institute held a very abbreviated summer course of some thirty lectures on the same topic. One is reminded of the definition given in Freshman Physics of a gas, viz. that it expands to fill any size container. The converse, compression, may also lead to a liquid or solid and (to compound the metaphor) may lead to indigestion. It is nevertheless possible, in a single lecture, to give the flavor of the subject and an idea of its present scientific and technological status, plus a description of its (past and future) interaction with mathematics.

AMS(MOS) subject classification (1970). 76W05, 49D99, 35R20, 00-02.
[1]This work was supported by U.S. Energy Research and Development Administration under contract No. EY-76-C-02-3077.

The outline of this short course within a short course is:

Controlled Thermonuclear Research (CTR) and
Magnetic Fusion Energy (MFE)

1. The field (CTR) and the working substance (plasma).
2. Ultimate potential as an energy source.
3. Present status as a field of science and as a branch of technology.
4. Present status with regard to the power producing goal.
5. The role of mathematics (and mathematicians) in CTR.
6. The role of CTR in mathematics.
7. A few "typical" applications involving interesting new mathematical structures.

Very briefly, with regard to items (3) and (4), we are beginning to approach a state of maturity in attitude, if not quite yet in content. With regard to item (5), our thesis will be that mathematicians can play a surprisingly large role, considering that the physical field is still "wild and woolly", so much so that the laws of physics are still being discovered; and with regard to (6) we venture a prediction that mathematical structures arising in plasma physics will enter the mainstream of the next century's mathematical analysis, just as fluid dynamics and electromagnetics did in the last.

STATEMENT OF THE PROBLEM. Fusion involves the collision of certain light nucleii (e.g. isotopes of Hydrogen, typically Deuterium or Deuterium plus Tritium) which may fuse to produce other nucleii (such as neutrons and Helium) and energy. In contrast, fission involves the splitting of certain heavy nucleii (typically Uranium or Plutonium), also with a release of energy.* There are many exothermic fusion reactions. About these we note only that the one which seems to involve the most accessible operating conditions uses a 50/50 mixture of Deuterium and Tritium. Pure Deuterium as the fuel (a later development) would avoid the necessity of manufacturing and handling radioactive Tritium (which is not found naturally but is obtained by using the neutrons generated in the reaction). Further down the road are technologically more difficult "esoteric" reactions which involve no neutrons at all and would eliminate all problems of radioactivity and neutron activated structural damage.

Nature abounds in <u>operating fusion power plants</u> - viz. the Sun and most stars. It would not suffice to merely reproduce a portion of the Sun in the laboratory. The reaction must be accelerated from a burning time scale of billions of years to one involving seconds, at most. At the opposite extreme, the time scale of an uncontrolled, very rapid reaction, as in an H-bomb, is too short to be a useful source of energy unless it is slowed down. Elementary estimates turn out to require ignition temperature of several times 10^8 degrees. The usual energy unit is the ev (one

* See accompanying article by M. Plesset.

electron volt $\sim 10^4$ °C) or kev ($\sim 10^7$ °C). We are interested in temperatures of 10 to 100 kev. The situation is entirely different from that in accelerators where the directed (not thermal) energy may be much higher; in a plasma we are concerned with a true kinetic temperature in which the velocity is more-or-less random and isotropic and is frequently close to Maxwellian. We say kinetic rather than thermodynamic temperature because the plasmas of interest will not be in equilibrium with regard to radiation. Radiation losses must be replaced by heat input or by the thermonuclear reaction itself; it is the latter that indicates an ignition temperature somewhere in the range given above. For an optimal temperature, ignition is conventionally given by the Lawson criterion, $n\tau > 10^{14}$, where n is the plasma density (ions per cc.) and τ is the lifetime of an ion (secs.). For reasons that will become apparent, any unmodified figure of merit of this type is more appropriate for use in Sunday supplements than as a scientific or technological measure of past progress or of the remaining distance from the goal, or as a comparison of the relative status of two experiments.

The time required for a significant burn depends on the plasma state but, for practical reasons, must range from an appreciable fraction of a second to several seconds. At 10 kev the speed of an electron is about 1/7 of the speed of light and that of a Deuterium ion, about 1/400. In an appreciable fraction of a second even the ions will traverse (or circulate around) the apparatus many thousands to millions of times. Although an accelerator (even more so a storage ring) confines particles for many more transits, this is done with very specialized techniques appropriate to particles in an extremely small part of phase space. The thermonuclear plasma, a true gas, requires confinement in a "bottle" which encompasses approximately all of phase space.

Turning to the density of a thermonuclear plasma, it is easily estimated from the energy density, or pressure. This must be comparable to that of steam in a conventional power plant, and efficiency evidently favors as high an energy density as is practicable. The pressure will be from one to several hundred atmospheres; the upper limit is determined by the strength of the structure. Given the temperature and pressure, we deduce that the plasma density is 10^{-2} to 10^{-6} of atmospheric - low, but by no means a good vacuum. Densities of actual plasmas of thermonuclear interest range from 10^{13} to 10^{17} particles per c.c. (atmospheric density is 3×10^{19}).

To confine a hot, dense plasma and keep it from physical walls requires a field of force. Gravitation evidently works! But the minimum size would not be very much smaller than the Sun. The only other relevant force field is the magnetic field. A convenient conversion factor is that one atmosphere equals 5000 gauss = 5 tesla (and $p \sim B^2$). Fields over 10^5 gauss (6000 psi) are not likely to be used in any large permanent structure (megagauss fields are usually only obtained explosively).

An evident complication is that magnetic fields (cf. the Maxwell stress tensor) exert highly anisotropic forces. This will be an important factor in the design of magnetic "bottles", and it also suggests that plasmas in contact with a magnetic field may frequently be anisotropic.

There is another confinement possibility which has been given the unusual term "inertial" confinement.* Instead of confining the plasma (force per volume = ∇p) by a magnetic field (Lorentz force = $J \times B$), it is confined by the "inertial force", $-\rho\, du/dt$. In other words it is not confined at all and is allowed to expand freely. A very small, very high density explosion, ignited by concentrating a high intensity laser beam or electron beam on a solid pellet, will be sufficiently attenuated by the time it expands to the size of the physical container that it does no significant damage (a conventional thermonuclear bomb could be so used in a large piston, say a mile in bore).

To be more precise, even with successful magnetic confinement, all the plasma and most of its energy eventually reaches a physical wall. The purpose of magnetic confinement is not to prevent but to slow down and control the loss processes, to deposit the plasma and its energy where it will do most good and least harm. More specifically, it is to assure that most particles which reach the walls (both charged and neutral atoms) are relatively cool. Part of the charged particle energy may be recovered directly as electrical energy (at high efficiency) by expanding the plasma against the confining magnetic field; another part, escaping at "open ends" (see below, magnetic mirrors) may have some of its streaming energy also converted directly into electricity; but most of the charged particle energy eventually reaches a physical wall where the energy is transferred to circulating coolant. Most neutrons deposit their energy in a neutron blanket surrounding the plasma, also transferring the energy to coolant. Some of what was originally charged particle energy reaches the walls in the form of high energy neutrals (through an intermediate charge exchange collision). Radiation is distributed rather widely in all the surrounding structure. Harmful effects (other than too rapid loss of plasma energy) are primarily dislodgement of impurities by energetic particles bombarding the walls, and structural damage of walls and coils by neutrons. Magnetic fields may serve almost as much to keep impurities out as to keep plasma in. In other words, the magnetic field is not so much a bottle as a porous plasma retardant. Nevertheless, many of the most useful theoretical models take the more ideal formulation.

The standard, popular (but correct) remark is that the fuel is inexhaustible and essentially free. There is enough Deuterium in a gallon of water (e.g. sea-water) to provide the energy equivalent of 300 gallons of gasoline, and the cost of extraction of the Deuterium is already negligible. What this means, of course, is that the cost to the consumer will be determined by the capital cost of a power plant over its lifetime,

* See accompanying article by F. Tappert.

distribution costs, and whatever ecological costs there may be in comparison with the competitive cost, at some future time, of alternative energy sources.

The description of the fuel as "inexhaustible" is sometimes contested because Tritium, which is an artificial element, is generated by neutron bombardment in a Lithium blanket surrounding the proposed D-T reactor, and it is estimated that there is only a hundred year visible supply of Lithium (this, incidentally, is comparable to the resource lifetime claimed for fission reactors). However, since substantial use of Lithium for this purpose would commence some 25 to 50 years in the future, it seems safe to assume that 125 years of technological progress (cf. space flight, radar, solid state in 1850) will be adequate to develop Tritiumless (D-D) and even neutronless power plants (using more esoteric nuclear reactions) in less than 100 years following the first operating D-T power plant.

The comparison between fission and fusion power with regard to economics, safety, time scale, ecological factors, etc. has probably already generated enough heat to fuel a medium-sized generator. This question would take us too far afield; let us just take as an exhibit, Fig. 1, courtesy of Stephen O. Dean, Assistant Director for Confinement Systems, Division of Magnetic Fusion Energy, U.S. Energy Research and Development Administration.

ADVANTAGES OF FUSION POWER:

1. FUEL SUPPLY PLENTIFUL, COST LOW
2. NO COMBUSTION PRODUCTS, VERY LITTLE RADIOACTIVE WASTE
3. LOW RADIOACTIVITY, REDUCED ASSOCIATED DANGERS
4. NO CHANCE OF RUNAWAY
5. NO DIVERSION OF WEAPONS - GRADE MATERIALS
6. HIGH EFFICIENCIES, POSSIBILITY OF DIRECT CONVERSION

Figure 1

The basic conclusion is that a hypothetical fusion reactor does have inexhaustible free fuel, is almost certainly much cleaner and safer from the point of view of radioactivity and other ecological hazards and suffers from only one clearcut disadvantage compared to coal, oil, or fission - it does not yet exist.

References: For an excellent historical account of Fusion up to 1958 see [1]; for a more up-to-date survey see [11].

PRESENT STATUS AND FUTURE PROJECTIONS. As an introduction to the assessment of fusion as a field of science or technology it is necessary to evaluate the magnitude of the problem, in particular its complexity. A proper appreciation of this point dominates all scientific and technological evaluations. Every once in a while there is a newspaper article which briefly assesses the status of magnetic fusion by stating that the problem has been worked on for n years "without success". This is extremely misleading. In fact, the state of the art has advanced by many orders of magnitude, with innumerable "breakthroughs" each sufficient to solve an ordinary difficult problem. But the path ahead will also require quite a few additional scientific and technological "breakthroughs". There is a very difficult public relations problem in simultaneously "pointing with pride" to the past and realistically evaluating the distant future.

A very brief statement serves to illustrate the magnitude of the technological requirements. An eventual (first generation) power producing reactor, as it is visualized today, would have a plasma core a few meters in minor diameter at a temperature in excess of 10^8 ^oC at a pressure of, say ten atmospheres, carrying a current of 10^7 amperes. Something like one meter distant from the plasma are coils at room temperature (or, more likely, superconducting, at approximately absolute zero). The entire structure, under large, complex stresses, is bathed in x-rays and other types of radiation and also in neutrons. There is provision to replenish the plasma (continuously or intermittently, as in an internal combustion engine), to remove reaction products and the heat of reaction, with remote handling for possible repair and periodic replacement of radioactive structures. Even without detailed analysis, the engineering problem is, to say the least, formidable.

The theoretical problem is at least as formidable. A simple (and very crude) measure of the richness of physical phenomena to be expected within a field is given by counting dimensionless parameters. For example, fluid dynamics is held to be spanned by two parameters, Mach number and Reynolds number (this is misleading - but not relevant for the present argument). A similar count in the simplest plasma (fully ionized, one species of ion) yields seven or eight instead of two. To be more specific, in order to eliminate the geometry, consider an infinite domain without boundaries and the classical steady, plane shock transition. In a gas, there is one parameter - the shock strength (this determines the thickness and all profiles, such as density, temperature, and the entire molecular distribution function). The equivalent problem for a fully ionized plasma in a magnetic field (uniform at infinity) has six dimensionless parameters. The large variety of qualitatively different physical mechanisms which govern the shock thickness fill a six-dimensional phase space; (the shock thickness is almost never simply related to a mean free path).

In Fig. 2 the various regimes of classical fluid dynamics and kinetic theory are located on a two-dimensional figure with axes M (Mach number) and 1/Re (Reynolds number). Qualitatively, the value of M measures the importance of compressibility and Re that of viscosity (including turbulence). Rarefaction effects (finite mean free path) are not included in fluid dynamics and are physically and intuitively distinct from compressive and viscous effects, but they appear on the same diagram, since the Knudsen number, K (mean free path divided by representative length) is given by K = M/Re. In other words, classical continuum fluid dynamics is more properly described as a one-dimensional parameter space, the two semi-axes where $K \to 0$ (one parametrized by M, the other by Re).

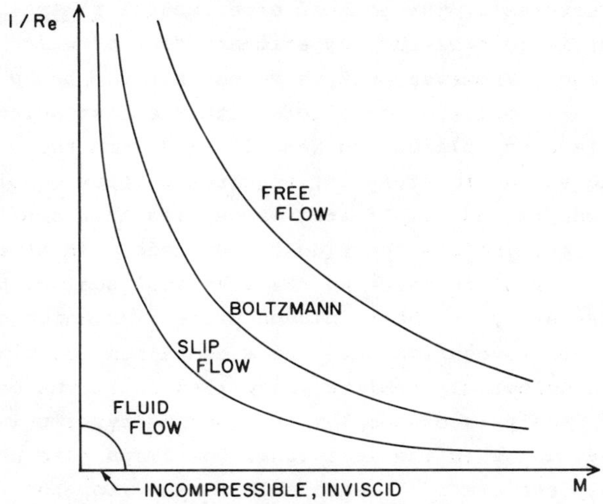

Figure 2: Regions of Gas Flow

Classical incompressible inviscid fluid flow is represented by the neighborhood of the origin in Fig. 2. One might perhaps describe the content of this theory in a single volume. Compressible theory would require a shelf of books (lining the M-axis) for comparable coverage, similarly for viscous flow, a row of books lining the 1/Re axis. A bookcase-full (not presently available) would be needed to provide classical kinetic theory of gases with similar coverage. An equivalent degree of coverage of all plasma physics would (perish forbid!) require the entire N. Y. Public Library devoted exclusively to this subject (and the world's GNP for many years to amass the knowledge in these books).

Restricting the field to plasmas of thermonuclear interest does cut down the magnitude of parameter space and of physical phenomena, but not enough to hope to blanket the area, either theoretically or experimentally, by any expenditure of effort and funds (e.g. magnetic fusion plasmas allow a density range of 10^4). Inevitably one must be extremely selective in choosing areas of study, and one must be constantly aware that the route followed is almost certainly not the best one. Nevertheless, whatever routes are followed <u>must</u> be pursued redundantly, with overlapping experimental parameters. Time does not permit elaboration of this essential point.

There is a further peculiarity that characterizes thermonuclear plasma physics in contrast to other areas of plasma physics and of physics in general. A plasma is a new substance, a "fourth" state of matter. What we must do is to learn its properties such as equations of state, transport coefficients, wave propagation characteristics, etc.; we must also learn to manipulate it - heat, cool, compress, pump, confine, purify - just as we do air or steam (or a new superfluid). Although nature is always exceedingly complex, physics gains it power (as does mathematics) by rejecting the complexities that occur in nature in order to study isolated "basic" phenomena (or structures). The goal of experimental plasma physics (as distinct from CTR) is to construct experiments each of which isolates an individual phenomenon. However, before we can study a hot plasma we have to catch one. This stipulation conflicts with the best scientific procedures. To create a hot plasma and keep it away from the walls, hot and undefiled for long enough to study it, requires complex combinations of experimental procedures and complicated geometries that conflict with the desire to simplify and isolate individual phenomena. In other words, one must be fairly successful in reaching the practical goal of plasma confinement before one can study the basic plasma properties which should be known in order to learn how to confine it. For a contained hot plasma the scientific and technological problems are presented all at once rather than in sequence. Analysis of complex interacting systems is commonplace in engineering, but not when the individual phenomena have not themselves been scientifically explored. A reasonable analogy for the fusion energy problem would be a mission-oriented program to land a man on the moon where, as part of the project, one must at the same time discover (and exploit) Newton's laws of motion, Maxwell's equations, electronics, and solid state physics. In addition, this is expected to follow a program plan which synchronizes and coordinates delivery of condenser banks with the planned date of the discovery of the magnetic monopole.

To assess the status of this complicated combination of mission-oriented basic physics and brand new technology we look at two questions:

1. To what extent are the relevant laws of physics known (in a practical sense such as, will turning a knob in one direction improve or destroy confinement)?
2. Is the state of the art (experimental or theoretical or numerical) such that any level of effort at the present time is sufficient to answer a specific question, or must we wait a decade or a generation for the state of the art to reach an appropriate level?

With regard to the second question, it is clear that the space program had to wait for solid state technology to provide an appropriate level of miniaturization, also for computers to be able to calculate course corrections during flight. This state-of-the-art requirement applies equally well to theory. Quantum mechanics flourished in its early days because the mathematical foundations of spectral theory already existed; relativity progressed rapidly because differential geometry was available beforehand.

Returning to CTR we point out that two Tokamak experiments, the TTT and the Alcator (for the present, these can be taken as undefined terms) reached some sort of terminal state during 1976. The first was a disaster; the second succeeded beyond all expectations. The discrepancy cannot be attributed to any significant difference in scientific ability, administrative planning, or funding hardship in the two organizations. The reason was simply that the relevant laws of physics were not sufficiently known beforehand (or even yet); there was (and is) no sufficiently refined theoretical plus computational model able to discriminate between the two disparate behaviors. In order to discover the answers, both experiments had to be performed.

Another fact (which requires too much space to document here) is that virtually every major confinement experiment to date has turned out to be <u>dominated</u> by a physical phenomenon which was not considered during its planning and design (this fact may be contested by some experts; it is nevertheless uncontestable).

It is clear that the requisite practical understanding of the physical (or even empirical scaling) laws is not yet known. What about the more serious practical question of the state of the art? This can only be answered by looking backwards. As an experimental example, the decade of the sixties is sometimes considered to have been fairly stagnant with moderate increases in $n\tau$ and temperature and numerous disappointments. But it was a time of revolutionary advances in plasma diagnostics and measurements. In 1960 if an experimenter was asked, what was the density or temperature of his plasma, one got a number in return (and, if pressed, a number surrounded by caveats). [I recall having made the statement, in the late fifties, that if an operating thermonuclear plasma were presented to us by an extraterrestrial visitor from a superior civilization, we would have no means of confirming whether it indeed was what it was represented to be.] Ten years later, in 1970, one could expect, in reply, to receive a flood of data giving profiles, density or temperature as a function of plasma radius and of time; in a few cases one would be overwhelmed with a complete distribution function in physical and velocity space. One cannot claim to have a science of plasma physics until plasma properties can be measured. One cannot begin to compare theory with experiment until detailed (profile) measurements can be made; the stability of a plasma, for example, will depend sensitively on the detailed profiles - not on just a few global properties. With a few exceptions, there was no way to compare theory with experiment until diagnostics reached a certain level of maturity (though, of course, very many such comparisons were made). There are still important gaps, e.g. in the measurement of current density profiles (essential to compare with most stability theories), and it would be most useful for comparison with theory to have localized measurements of the fluctuation spectrum.

A similar case can be made with regard to the state of the art in MHD theory of equilibrium, stability, and diffusion. In each of these, the degree of theoretical sophistication and professionalism which (with a very few exceptions) is absolutely necessary for purposes of experimental verification has only begun to be achieved since about 1970; briefly, for stability, global eigenfunctions rather than local criteria (nonlinear results are still in infancy); for equilibrium, nonlinear and topological evolution; for diffusion, time-dependent nonlinear evolution; in each case, with a very careful balance between analytical and numerical inputs.

To summarize the state of the art in CTR after 20 years of research and development, consider as an analogue the sequence:

Kitty Hawk
Lindbergh
Commercial Jet
Vanguard Grapefruit
Appollo
Self-sufficient Space Station - Population 1000

I would estimate that CTR has reached the Commercial Jet stage on the way to an eventual self-sufficient space station. Unfortunately, except for a few technical applications, there seem to be no significant commercially useful intermediate stages on the way to a fusion power plant.

One fault in evaluation which has caused untold misunderstanding of progress in the field is the "one-clever-breakthrough syndrome" which is propagated equally by scientists (usually outside the field) and by newspaper reporters. Progress to date has been measured in orders of magnitude (and in scores of breakthroughs); but the path ahead is also measured in orders of magnitude. To counterbalance this is the fact that the payoff is also valued beyond measure, provided that the end result is not exorbitant in cost.

Another fruitless evaluation is the comparison with the stages in the evolution of the fusion reactor. The difficulties are noncomparable and occur in a different sequence. The relevant laws of physics in the fusion problem were contained in the nuclear reactions and their cross-sections. This is a triviality for the fusion problem, but there is no fission analogue to plasma turbulence, instability, or any cooperative effects in the neutron "fluid". The chemical problems of separating U_{235} and manufacturing UF_6 bricks have no analogue in plasma physics. The actual construction and performance of the Stagg Field fission experiment (given the bricks) could be done by appropriately coached high school students, whereas to just carry out a contemporary (far from reactor stage) plasma confinement experiment, given all the equipment, (including, say, Thompson scattering or laser holography), requires a highly trained professional team of experts linked to a complex, real time computing system.

In any discussion of fusion someone is sure to ask, when will scientific feasibility, or a demonstration reactor, or commercial fusion power be reached? In 1962 the following reply was given:

1) The theoretical problems are extremely difficult.
2) The experimental and technological problems are even more difficult.
3) Guessing when the problem (in any phase - scientific feasibility, prototype reactor, commercial power) will be solved is most difficult of all.

There is no reason to change this reply after a lapse of fifteen years. Several orders of magnitude still separate the present from where we wish to be. In covering this territory it is certain that new, unexpected physical phenomena will be encountered; new problems not yet formulated, will have to be solved; new laws of physics will be formulated. One can, with some risk, predict technological advances - but no scientist has (or should have) the temerity to foretell when new physical laws will be discovered, or how long it will take to solve a problem before the problem has surfaced. Such estimates, when they are necessary, are the proper province of scientific administrators, and where social issues are involved, elected representatives. But when they are made by scientists, they should not be confused with scientific judgments.

In a situation where predictions are so difficult (demonstrably so, historically), why are they so plentiful? There appears to be an equivalent of the axiom of business, that where there is a demand, it will be supplied. The same can be said of theoretical "explanations" of extremely complex experimental situations and predictions of the unpredictable. There seems also to be an application of a form of Gresham's Law, that poor science will drive out good. The pressure on theory to agree with experiments and on experiments to agree with schedules is unbearable. If an administrator asks for a prediction, or an experiment demands an explanation a definite answer is always more acceptable than an equivocation. There has been recent pressure (in other fields) for "one-handed" scientists, but it would seem that a scientist who has lost his "on-the-other" hand has ceased to be a scientist (even if, on-the-other hand, he turns out to be more useful to society by choosing not to be a scientist).

There is a final and rather subtle point concerning predictions using complicated mathematical models in complex physical situations. In yesterday's lecture* we were presented with some extremely complex economic situations. Social scientists have one advantage in almost never having succeeded in formulating a mathematically precise model of any part of their world; they (usually) retain a certain amount of cynicism with regard to mathematical models. "Hard" scientists are accustomed to accepting extremely accurate mathematical models as a norm; the fact is that, practically speaking, physics is <u>defined</u> to be that part of the universe

---

* See accompanying article by G. Dantzig.

which is accurately describable by mathematical models (e.g. turbulence has almost been banished from the subject). The fact that one does not predict a real airplane's performance from a single formula, but rather builds an onion of successive layers of partial theories and specialized experiments is beginning to bear analogues in plasma confinement theory.

All this talk of complexity may sound like pessimism or a cause for indecision. The opposite is in fact true. There are many examples (of which we cite only a few) where the fact of complexity and lack of accurate knowledge gives the correct, positive course of action. For example, should we concentrate on what seems to be the single best approach? The answer is clearly, no, it is much too soon. The laws of Physics, and the empirical scaling laws change with each experiment. Should we build one large experiment at $N to test one hypothesis or four smaller ones at $N/4 each? The better answer, for most rapid progress toward a reactor is obvious (build four), even though this runs counter to actual administrative practice. Since not all avenues can be tried, the optimum procedure would seem to lie somewhere between a one-shot gamble and a cross-the-board, safe approach. Of course, a relevant social or political input may prevent the question from ever taking the above form. There is also an interesting psychological point in the use of the phrase "real world" - are the laws of physics or of society and the budgetary world more "real", and which laws are more dangerous to flout?

The physical complexity is actually the single most important reason for a belief in the ultimate success of the program. Whenever one direction proves disappointing, there are ten alternatives to try. All the leading contenders today defy Murphy's Law (if anything can go wrong, it will) - each one behaves significantly better than one had any reasonable right to expect when it was planned. And the fact that so far almost all experiments end up heading in directions that were not planned, is evidence of the intelligent use of the flexibility inherent in the variety of options.

References: For a fuller account of some of the points in this section, and a survey of plasma physics in general see [7]; for a programmatic survey of future plans, see [19]; for technical studies of reactor possibilities, see [16] and [17].

BASIC CONFIGURATIONS. We now describe some of the basic magnetic confinement configurations, together with a glossary of common terms used to describe them. Some of the names are functional or descriptive; others have lost their original meaning with the passage of time.

A prototypical confinement geometry (Fig. 3a) is an infinite plasma cylinder surrounded by a vacuum magnetic field, terminated at a conducting wall (which carries currents that can be considered to be the source of the

the field). In Fig. 3b the field is axial, B_z, and the current is azimuthal, J_θ; this is called a θ-pinch. The dual configuration, in Fig. 3c, with azimuthal field B_θ and axial current J_z is called a z-pinch. The name pinch is historical, referring to a transient phase; an axial electrical discharge between two electrodes gives rise to a distributed z-current, the constituent elements of which attract one another, "pinching" until the electromagnetic attraction is balanced by the compressively increasing plasma pressure. The conceptually simplest pinch configurations contain a constant pressure plasma but no field within the cylinder and a vacuum (harmonic) magnetic field outside (B_z = const., B_θ = const./r).

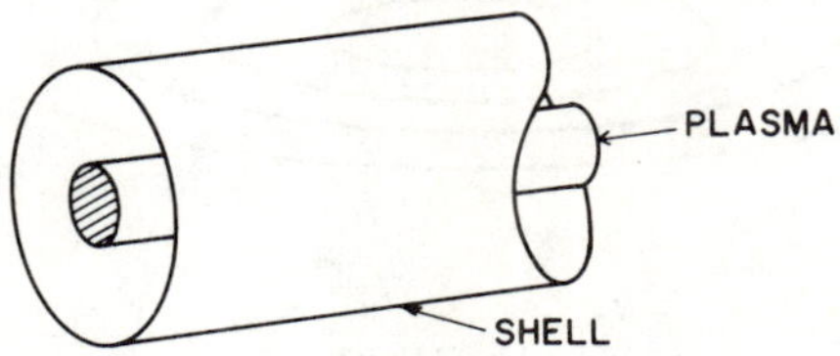

(a) Plasma and Wall

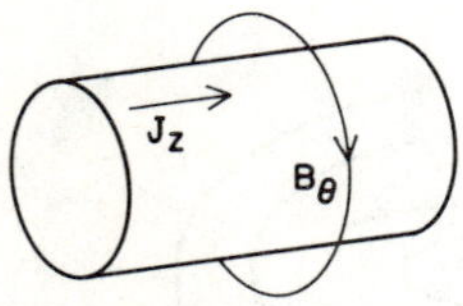

(b) θ-Pinch

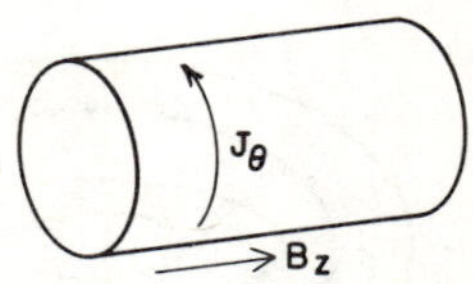

(c) Z-Pinch

Figure 3: Cylindrical Equilibria

For reasons of stability, the z-pinch is always accompanied by an axial B_z, taken to be uniform in the simplest case, with equal values inside and outside the plasma cylinder. The combination B_z inside, a different value of B_z plus B_θ outside, both J_θ and J_z on the interface, is called a screw pinch. The pressure balance condition at the interface is that $p + \frac{1}{2}B^2$ be continuous, or that $p = \frac{1}{2}B^2_{out} - \frac{1}{2}B^2_{in}$ $(B^2 = B^2_\theta + B^2_z)$.

A more general cylindrical equilibrium, also termed screw pinch, has a distributed pressure profile p(r) together with fields $B_\theta(r)$, $B_z(r)$ and currents $J_z(r)$, $J_\theta(r)$. Ampere's law, J = curl B or

$$J_\theta = B'_z \quad , \quad J_z = -(rB_r)'/r$$

coupled with the pressure balance, $\nabla p = J \times B$ or

$$\frac{d}{dr}\left(p + \frac{1}{2}B^2_\theta + \frac{1}{2}B^2_z\right) + B^2_\theta/r = 0$$

allow two of the profiles p, B_θ, B_z, J_θ, J_z to be given arbitrarily. The plasma is confined (two-dimensionally) if p drops to zero, p(a) = 0, and the surrounding field, r > a, is harmonic. With this boundary condition for p, |B| is continuous at the interface; there can be a surface current, but it is usually set equal to zero (B_θ and B_z continuous).

Other common plasma cross-sections are the elongated Belt-Pinch (Fig. 4a), the Doublet (Fig. 4b) with topologically non-simple flux surfaces, and the Divertor (Fig. 4c; the two indicated coils are outside the plasma leaving a simple magnetic topology inside the D-shaped plasma).

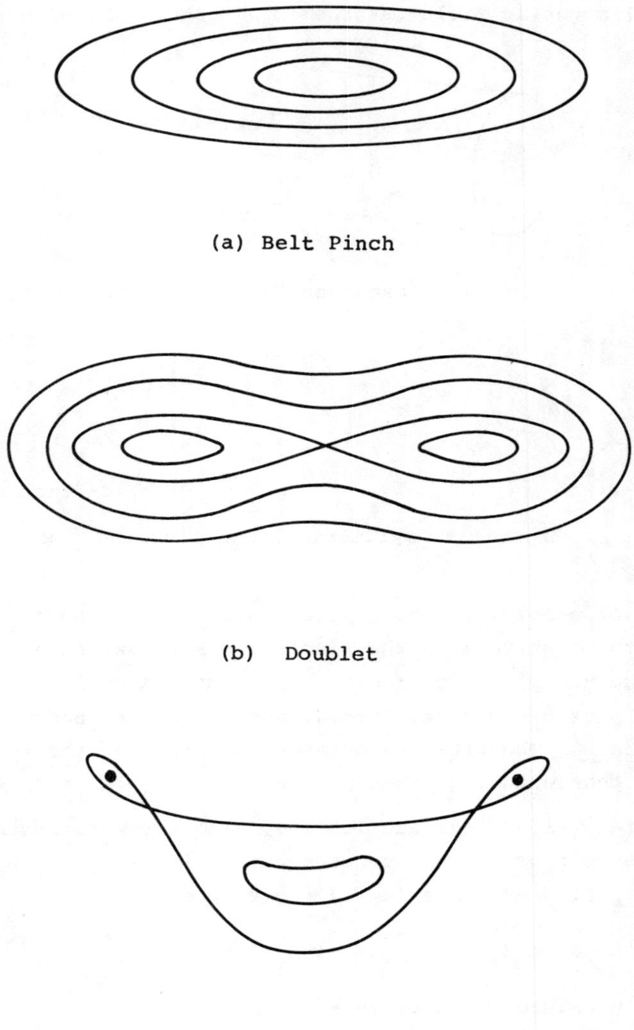

(a) Belt Pinch

(b) Doublet

(c) Divertor

Figure 4: Noncircular Cross-sections

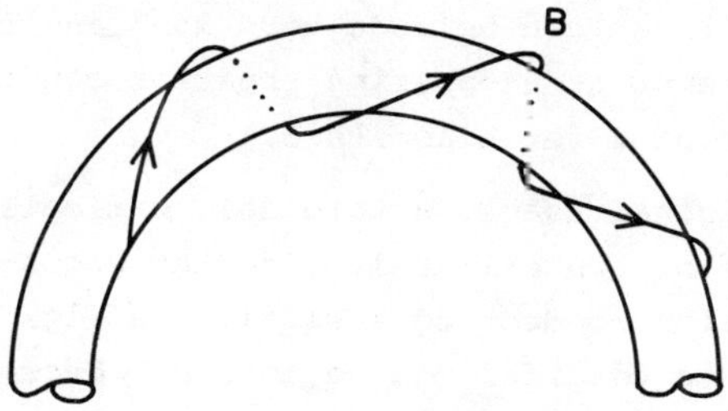

(a) Closed System

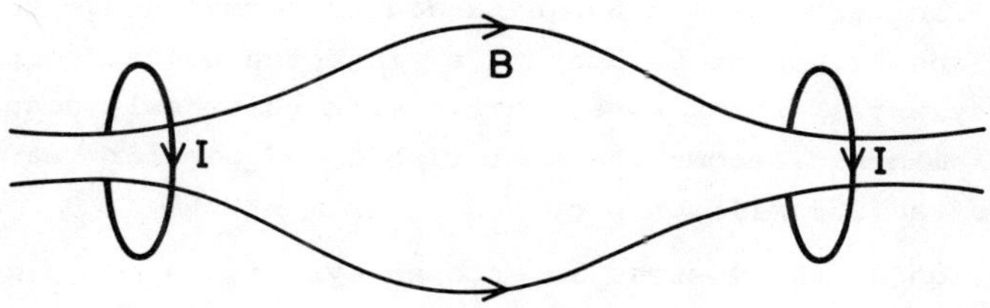

(b) Open System (Mirrors)

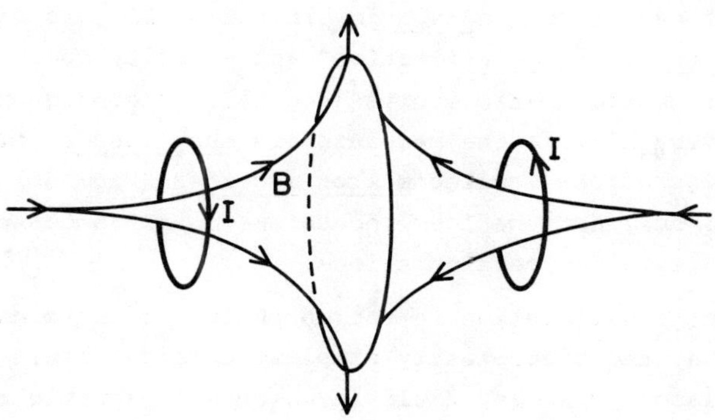

(c) Cusped Geometry

Figure 5: Three-Dimensional Confinement

To confine (or approximately confine) the plasma three-dimensionally, the (θ, z, or screw)-pinch can be bent into a torus, eliminating the open ends (Fig. 5a); or the ends can be squeezed to reduce the end losses (Fig. 5b); or the cylinder can be made very long and the end-flow simply ignored; with no attempt at stoppering, reactor conditions would require a length on the order of a few kilometers.

According to original usage, a toroidal, ohmically heated discharge (generic screw-pinch) of approximately circular cross-section, enclosed by a circular cross-section conducting shell (as in Figs. 3a, 5a) with poloidal field small compared to toroidal (large toroidal current), low $\beta \equiv p/B^2$ (plasma energy small compared to magnetic energy), and the "safety factor" $q = RB_\theta/aB_z$ (reciprocal of the magnetic rotation number, R and a are major and minor plasma radii) on the order of two or three, is called a <u>Tokamak</u> (Russian acronym). Since the plasma has finite resistivity, toroidal current implies $\oint E \cdot dx \neq 0$. The magnetic flux threading the torus (usually in a transformer core) must vary in time; thus this is an inherently transient system. The term Tokamak has expanded in common usage to include all plasma shapes and values of β, heating by injected beams, compression, radio-frequency waves, shock waves, etc.; with this newly acquired flexibility the Tokamak has become the most viable and popular reactor concept (to the tune of a $250 million proposed experiment).

The question of end losses, in an open system, is not simple. There are at least three qualitatively different mechanisms, depending on parameters. For small mean-free-path (not usually relevant with reactor parameters), the flow is sonic at an end, or throat (but this is complicated by the fact that, in MHD as distinguished from ordinary fluid dynamics, there are three characteristic speeds and three possible sonic transitions). For large mean-free-path and <u>adiabatic</u> orbits (small gyro radius), the loss mechanism is <u>magnetic mirror</u> reflection* and velocity space diffusion into a <u>loss cone</u>; for non-adiabatic orbits (generally a consequence of locally weak fields or very high β) the mechanism is <u>cusp losses</u>. There are corresponding configurations, called <u>mirror machines</u> (Fig. 5b) and <u>cusped geometries</u> (Fig. 5c), but the loss mechanisms do not necessarily apply to the correspondingly named configurations.

An important consideration in the morphology of toroidal confinement geometries is that the conceptually simplest toroidal field configuration does not hold plasma, viz. any field in which all magnetic lines are circles with a common axis. A simple macroscopic argument shows that any solution of $\nabla p = J \times B$ with the stated symmetry must have $p = p(r)$ [in cylindrical coordinates (r, θ, z)], which states that p is constant on an unconfined infinite cylinder; alternatively, any (transiently) confined <u>plasma will have a net, unbalanced (out</u>ward) force. A somewhat similar

* An asymptotic orbit expansion indicates that the <u>magnetic moment</u>, $\mu = \frac{1}{2}mv_\perp^2/B$, is approximately constant. Since $\frac{1}{2}mv_\perp^2 + \frac{1}{2}mv_{||}^2$ is constant, a large value of B is not accessible, causing reflection of $v_{||}$ between mirrors.

miscroscopic orbit argument shows that nonconfinement follows from oppositely directed ion and electron orbit drifts.

There are several common complications of this unworkable simple symmetry which are introduced in order to provide confinement. One goal, starting from the microscopic orbit model, is to create a family of confined flux surfaces which (to a certain approximation in orbit theory) guide particle orbits. In the Tokamak, this is done by inducing a toroidal plasma current. In the Stellarator, nested flux surfaces are produced in the vacuum field by bending the configuration into a Figure-Eight, or by applying suitable external windings, usually helical, to give a nonzero magnetic rotation number. A different solution (with closed magnetic lines) is to provide a series of bumps (as in a string of mirror machines) along the toroidal direction, maintaining closed but noncircular field lines; this is the EBT (Elmo Bumpy Torus).

From the point of view of the momentum balance, $\nabla p = J \times B$, one is led to consider external coil systems which counterbalance the net outward force on the plasma in a purely toroidal field; this is the Scyllac. One practical set of Scyllac coils is helical, as in a Stellarator, but the optimal solutions (as well as simple physical intuition) for the traditionally low β Stellarator, based on vacuum flux surfaces, are quite different than for the high β Scyllac, orginating as a θ-pinch, and based on force balance considerations.

It is interesting to point out that the original Stellarator concept was the consequence of a nonexistence theorem (under too much symmetry). Later, deeper analysis showed that all Stellarator "solutions" remain subject to a more sophisticated lack of existence of static equilibrium, leading in practical cases to an irreducible fluctuation level and essential complications in the topology (cf. Refs. [6], [8]).

The advantages and disadvantages of the various magnetic configurations coupled with the many ways of creating and maintaining plasma density and temperature are not at all evident, and, despite complex and sophisticated theoretical calculations and well-documented experiments are, to a considerable extent black (or at least gray) magic. For example, a simple mirror machine created by radio frequency heating is observed to be stable; but if it is created by plasma injection it requires special stabilizing windings (Ioffe Bars, or cusped windings, Fig. 5c). Another example is the Tokamak which requires for its operation an initial "anomaly" factor of some 200 in resistivity, dropping almost immediately to close to unity. Without the large initial anomaly it would be an uninteresting cold plasma. Without the precipitous drop, the plasma losses would be intolerable.

State of the art confinement experiments at the present time are in the \$10-\$20 million range; in the next five or ten years they will reach the \$50-\$250 million range. The upper figure would perhaps produce an experiment with significant burning if filled with Tritium as well as Deuterium.

The present magnetic confinement budget is approximately \$220 million for FY 77.

References: No single source is complete, but additional information with regard to alternative configurations will be found in [1] and [11].

MATHEMATICS AND MAGNETIC CONFINEMENT RESEARCH. It is traditionally felt that mathematicians contribute significantly only to a mature scientific discipline after the dust has settled and the physical laws are well established. We wish to demonstrate exactly the opposite. Because of (not despite) the great complexity and the difficulty of performing experiments in magnetic confinement plasma physics, mathematics (and mathematicians) not only can but must play a vital role in formulating the basic framework for future development and exploitation of the field.

One could write down what many scientists would believe is a moderately accurate system of governing equations for a thermonuclear plasma. This would contain several coupled Boltzmann equations for ions, electrons, neutrals, α-particles and a large number of ionized species (such as multiply ionized oxygen, iron, etc.), together with Maxwell's equations and the coupling to various types of radiation, plus some model for the influx of impurities from the walls. Nonlinear effects, turbulence, and changes in magnetic topology are just a few factors which we know will play a very important role. To be quantitatively accurate, this model would have to include three space dimensions and time. Unfortunately, no such model could yield any useful information without drastic modification. Very crudely, one might assign a figure of merit to a theoretical model as the product of three factors, one being the faithfulness of the model to physical reality, the second to geometrical reality, the third being a measure of its mathematical tractability. To be precise, a model which is "solved approximately" should be described as a more elementary model which is <u>solved</u> (no quotes). It is clear that any useful model will have to set aside 90% of the physical complexity or (and?) 90% of the geometrical complexity. Ultimately, understanding of such a problem will come, not from one ingenious model, but from synthesis of very many interlocking ingenious models.

There are three popular misconceptions worth brief mention. The first is that the validity of a physical theory is determined primarily by experimental confirmation; the second is that, in order to be useful, a model must yield quantitatively accurate results; the third is that physical evidence and intuition provide the equations, after which mathematical analysis assists in obtaining the solutions.

Relevant to the first two points is the fluid dynamics model of incompressible, irrotational potential flow. This is a very poor quantitative model of a real fluid; a half-century ago it was obsolete (d'Alembert's paradox, etc.). Today, every fluid dynamicist spends a large amount of time developing a strong quantitative intuition about this non-existent fluid. The reason is twofold. First, it makes accessible an enormous mathematical literature (in real geometries). Second, it provides the basis for describing a real fluid in terms of its deviation from ideal behavior. The ideal model can become quantitatively accurate after it is modified in certain well-defined ways at boundary layers, wakes, and shocks. Even when not strictly quantitative, the combination of potential flow plus modifications is much more valuable for its physical insight than is the more accurate, full Navier-Stokes equations with their much more limited analytical theory and greater dependence on case by case numerical solutions. The value of potential theory is much greater today than it was a half-century ago, because today its theory is much more powerful and its limitations are quite precisely known.

A similar concept in plasma theory is the concept of frozen flux, that magnetic field is carried with the fluid (cf. vorticity in fluid dynamics),

$$\frac{\partial B}{\partial t} + \mathrm{curl}(B \times u) = 0$$

(B is the magnetic field and u the fluid velocity). This is very rarely an accurate, quantitative description of a plasma. But it is almost impossible to hold a rational discourse about a real plasma other than in terms of deviations from this ideal model.

How does one determine which 90% of the "real" problem to throw away? One consideration is mathematical solvability: (a) is the problem well-posed? (b) are powerful tools available? No physical evidence or intuition can ever have any relevance to the question of well-posedness. The answer to this question is known to depend on <u>arbitrarily small</u> changes in formulation. No arbitrarily small change can have physical (as distinguished from mathematical) meaning. The laws of physics are not the same as the laws of mathematics; equations obey the latter. What physical intuition <u>can</u> contribute is a strong psychological belief that <u>something similar</u> to a given formulation should be useful; this will encourage exploration of other neighboring systems if a particular one is mathematically unacceptable. What muddies this point in practice is that an approximate mathematical "solution" of an ill-posed problem may also be an approximate <u>solution</u> to an, in some sense, neighboring well-posed problem; (or this may be only a pious hope). For example, an asymptotic formula can be asymptotic to two different exact results; and an asymptotic procedure applied to an improper equation may give a formula which is correctly asymptotic to another (not even formulated) equation.

To return to the necessary amputation of most of the physics and geometry, intuition and experience are evidently crucial. But <u>mathematical</u> intuition and experience are at least as important as <u>physical</u> intuition and experience. How is it that in many fields scientists obtain accurate results by what are, mathematically speaking, very sloppy means? The answer is complex; but most crucial is a feedback teaching mechanism which slaps your hand when you stray too far from the proper path. In a field with frequent and accurate recourse to experiment, this provides a very effective interactive mechanism. In a field such as the one we are now describing, the experimental feedback is very weak and ambiguous; (the consequences are seen in the number of discarded models and calculations which litter the literature). However, since there is always a more accurate and more complicated <u>equation</u> which one would like to be able to solve, one can admit purely mathematical hand slaps as feedback, based on qualitative and semi-quantitative inputs from the more exact model. A careful evaluation will show that, up to now, mathematical care and mathematical intuition has played a large role in obtaining lasting results in confinement plasma physics.

"Mathematical care" must be carefully distinguished from mathematical rigor. A century of experience in applying mathematics to physical systems is available to offer reliable guidance as to what kinds of expansions, iterations, scalings, simplifications, and approximations are likely to work in differing mathematical contexts and which ones are likely to lead to a well posed formulation. In any complex problem on which a number of mathematically incompatible models have been brought to bear (e.g. microscopic and macroscopic, ideal and dissipative), interpretation and synthesis of the disparate results in relation to the desired (unsolved) metaproblem is much more valuable than the raw, unevaluated data. The physical significance of singular and nonuniform results is generally meaningless if interpreted literally and is invaluable if put into proper context. Some of this is susceptible to proof; much is not. In large part it consists of being sensitive to mathematical warning signals that a formal procedure is about to run into a snag; (this is analogous to the smell of burning insulation in an experiment). The misapplication of physical intuition (as derived from natural phenomena) to properties of equations is just as dangerous as is the misapplication of infinitely precise mathematical distinctions to nature (which latter error is, incidentally, the source of most "paradoxes" in physical science).

This type of <u>mathematical intuition</u>, used to guess at what is likely to produce a correct result, <u>is</u> mathematics even in the absence of proof since it concerns mathematical structures, not directly nature. In a sense, mathematicians have given up a part of their birthright in downgrading qualitative and intuitive analysis of mathematical structures (leaving this important task to non-specialists). Restricting the subject matter of mathematics to what has been proved is both illogical and unwise.

As an illustration, let me relate a (true) anecdote concerning two scientists, one an applied mathematician, the other an experimental physicist, both listening to a theoretical lecture which involved a lengthy formal expansion leading to a result which was presented without further critical comment. Quoth the physicist, "He may have done the mathematics correctly, but he left out the physics". Replied the mathematician, "He may have done the physics correctly" [since formal calculations without error estimates are usually done by physicists], "but he left out the mathematics". In any event, from the terminology employed, both observers agreed that the most important component had been omitted.

A brief word about numerical computation is in order since this will inevitably be an essential tool in any complex field. We have observed an empirical rule that a modest amount of appropriate analytic preparation will frequently be rewarded a thousand fold in accuracy or calculation time.

In summary, since the traditional corrective action of experiment is weaker than usual in this field, more than the usual reliance on logical consistency and mathematical caution is demanded. In particular, if the model to be used has already been formulated, it is usually too late to take in a mathematical consultant. We remark that all the preceding generalities can be related to concrete examples.

[End of Inspirational Message]

References: For mathematically oriented surveys of the field see [3], [4], [10], [13].

MATHEMATICAL EXAMPLES

(a) Introduction. We briefly describe a class of related problems which serve to introduce some interesting new mathematical structures. The physical problems have great current interest and the mathematical formulations seem likely to lead to significant mathematical developments. They concern new types of well posed problems which have required development of a body of theory and discovery of practical techniques of solution; on the other hand, they are not so radically unfamiliar (all too common in plasma physics) as to lose contact with the body of established mathematics. They are related to ordinary and partial differential equations but are different from both; they can be termed nonlinear "functional" equations, but this serves no purpose. They exhibit the classical thread that a physical situation motivates the mathematical problem, but it does not give a precise formulation nor does it suggest practical methods of solution.

The family of problems concerns adiabatic compression and resistive diffusion of a plasma. Implicit in the word adiabatic and in the concept of (slow) diffusion as distinguished from (fast) wave motion and flow, is the concept of a quasi-static one-parameter family of solutions of the equilibrium pressure balance,

$$\nabla p = J \times B \quad , \quad J = \text{curl } B \tag{1}$$

That we are concerned with a <u>family</u> of solutions of (1) comes from the (adiabatic) variation of external constraints such as coils, or through changes in internal pressure or current profiles induced by diffusion.

For simplicity (and almost by necessity)* we shall restrict ourselves to the two-dimensional form [(x,y), z ignorable]

$$\begin{cases} B = n \times \nabla\psi + n\, B_z \quad , \quad n = \nabla z \\ p = p(\psi) \quad , \quad B_z = f(\psi) \\ \Delta\psi = -\dot{p}(\psi) - f\dot{f}(\psi) \quad , \quad \cdot = d/d\psi \end{cases} \tag{2}$$

or the axially-symmetric form [(r,Θ,z), Θ ignorable]

$$\begin{cases} B = \nabla\Theta \times \nabla\psi + nB_\Theta \quad , \quad n = r\nabla\Theta \\ p = p(\psi) \quad , \quad rB_\Theta = f(\psi) \\ \Delta^*\psi \equiv r^2 \,\text{div}(\nabla\psi/r^2) = -r^2\dot{p} - f\dot{f} \end{cases} \tag{3}$$

We are looking for a family of solutions of (2) or (3) which depends continuously on a parameter. If the parameter is contained in <u>given</u> functions $p(\psi,t)$ and $f(\psi,t)$ or in a varying domain or boundary condition [with $p(\psi)$ and $f(\psi)$ fixed], we have a classical elliptic problem with known criteria for uniqueness, bifurcation, etc. If the problem is <u>adiabatic</u>, the two given profiles $p(\psi)$ and $f(\psi)$ are replaced by two adiabatic constraints (profiles) which serve to determine p and f. For <u>diffusion</u>, there are additional evolutionary equations with time derivatives to govern the changing p and f profiles. In neither the adiabatic nor the diffusion problem is the formulation of a well-posed problem evident.

Both of these problems are very singularly related to the dynamical equations of motion of an ideal or resistive magneto-fluid. In broad outline, the ideal equations, linearized about a static equilibrium [viz. about a solution of (1)], can be related to a formally self-adjoint operator, L, whose eigenvalues, $\lambda = -\omega^2$, are the natural frequencies of oscillation. L is a sixth order operator whose spectrum contains continua, accumulation points, dense sets of points, and more pathology. With regard to the original question concerning a family of equilibria depending on a parameter, the significant feature is that if the origin, $\lambda = 0$, belongs to the spectrum of L for a given solution of (1), then no neighboring, linearly perturbed equilibrium exists. A <u>point eigenvalue</u>, $\lambda = 0$, in the linearized

* With the exception of a handful of cases with special symmetry, the general plasma equilibrium (1) admits no well-posed formulation, cf. Ref. [6]. This fact demands careful interpretation of the literature of formal expansions, most of which do not have the requisite symmetry, Ref. [8].

problem is a signal that there is a bifurcation of solutions in the nonlinear problem. Plausible arguments and examples can be given to show that a continuum which extends to the origin is a signal that the flux surface topology of the nonlinear solution is about to change, but (when correctly formulated), without any loss of uniqueness. There is also the possibility of simultaneous bifurcation and islation or anislation (appearance or disappearance of islands) for a point eigenvalue $\lambda = 0$ at the end of a continuum.

Flux conservation, implying invariance of all topological properties of field lines and flux surfaces, is a property common to both the linear and nonlinear ideal fluid equations of motion; it would therefore seem to be appropriate to adopt it as a constraint for any model of adiabatic variation. In particular, the value of the flux, ψ, is constant at any critical point, $\nabla\psi = 0$. This topological constraint prevents continuation of both the linear and nonlinear adiabatic models past a point at which a continuum touches the origin. We are therefore led to generalize the definition of adiabaticity, dropping the invariance of ψ at a saddle point (the presumed mechanism is a dissipative boundary layer in the neighborhood of a separatrix). Simply allowing a change of topology would permit infinitely many continuations of the equilibrium. The situation is handled in a way which is analogous to the introduction of shocks in ideal fluid dynamics. Several (usually three) adiabatic regions which abut along a separatrix (cf. Fig. 6) are joined by jump conditions analogous to the

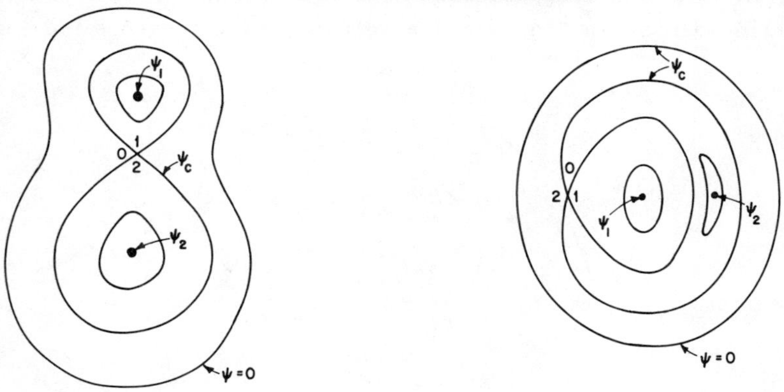

(a) Doublet Type (b) Tearing Type

Figure 6: Islated Domains

Hugoniot conditions across a shock wave. The matching conditions are suggested by appropriate conservation laws. This procedure turns out to specify the unique continuation of a generalized adiabatic process past a change in topology. As in shock theory, no properties of the dissipative mechanism need be known, so long as they are compatible with the hypothesis that the separatrix boundary layer is thin. Total flux is conserved but there is "breaking" and "reconnection" of magnetic lines; similarly, there is conservation of total mass, but redistribution among the regions.

Depending on whether appropriate regions grow or shrink, fluid and field can either mix or split; the two processes are evidently thermodynamically different. Problems of creation, growth, destruction, and interference of magnetic islands can be handled without overt reference to any dissipative mechanism. The evolutionary time scale is determined by the variation of the external constraints and not by a dissipative time scale.

The present state of this subject comprises a modest number of proved mathematical theorems, somewhat more in the way of formal boundary layer and other expansions and examination of singularities, and a large amount of empirical but accurate numerical confirmation of conjectured well-posed problems and of formal analytic singular and limiting behavior, the latter including, in particular, bifurcation, islation, and a host of boundary layers near the plasma edge, near a small emergent island and near the separatrix. The generalized adiabatic formulation with jumps has also been confirmed by numerical calculation of the resistive problem in the limit of small resistivity and, in a few cases, by comparison with plasma experiments.

If there is any point to this, one can also make an obvious connection with catastrophe theory.

(b) The Adiabatic Problem and Generalized Differential Equation in a Simple Geometry. After this qualitative introduction, we turn to a slightly more detailed exposition. To be precise, consider a simple two-dimensional domain as in Fig. 7 and a solution of (2) with simple nested

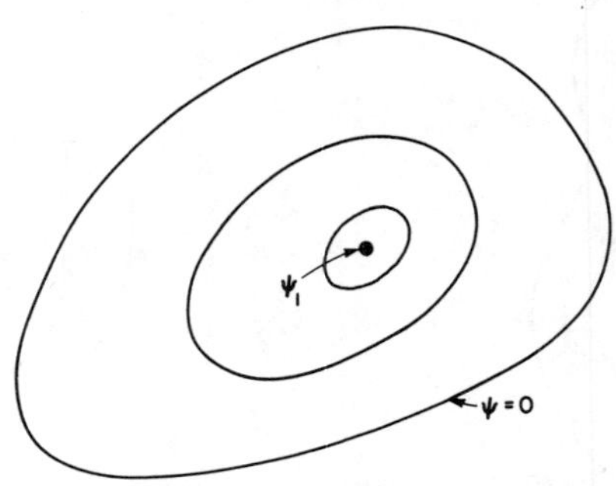

Figure 7: Simple Domain

contours ψ = const. as shown. For adiabatic constraints it is natural to introduce the volume $V(\psi)$ (area in 2-D) within a flux surface (contour) ψ = const. The adiabatic constraints take the form

$$(4) \qquad \begin{cases} p = \mu(\psi')^{\gamma} \quad , \quad ' = d/dV \; , \; (\dot{} = d/d\psi) \\ f = \nu\psi' \end{cases}$$

where $\mu(\psi)$ and $\nu(\psi)$ are adiabatically invariant (given) profiles to replace p and f. Eliminating p and f in favor of μ and ν in (2) yields the rather opaque functional equation

$$(5) \qquad \Delta\psi = -\dot{\mu}(\psi')^{\gamma} - \gamma\mu(\psi')^{\gamma-2}\,\psi'' - \nu\dot{\nu}(\psi')^2 - \nu^2\psi''$$

To gain understanding we simplify by setting $\gamma = 2$, $\nu = 0$, $\mu = 1/2$, and obtain

$$(6) \qquad \Delta\psi = -\psi''$$

or for greater generality

$$(7) \qquad \Delta\psi = F(V,\psi,\psi',\psi'') .$$

This equation has been called a QDE (for "queer" differential equation or from the sequence O, P, Q), or, more prosaically, a GDE (generalized DE).

Even the simplest form (6) is rather unusual. On the left is an elementary elliptic operator on $\psi(x,y)$; on the right is the equivalent 1-D elliptic operator on $\psi(V)$; but the 1-D independent variable V is defined in terms of the contours, ψ = const., of the 2-D dependent variable. The fact that the 2-D and 1-D differential operators are the same order will be seen to be crucial.

To return to more familiar territory, consider the elliptic problem with $p(\psi)$ and $f(\psi)$ given. The correct elliptic boundary condition is to specify ψ, e.g. $\psi = 0$ on the boundary. The value ψ_1 at the "center" (assuming that the topology happens to turn out as shown in Fig. 7) is determined together with the rest of the solution. If the shape of the domain is altered, the value of ψ_1 will change. The adiabatic statement of flux conservation for the entire domain demands that ψ_1 be fixed under domain variation. In other words, to be physically appealing, the systems (5) or (6) [and for mathematical reasons, also (7)] must allow specification of ψ_1 in addition to an elliptic boundary condition. This is not compatible with any known type of differential equation; (it might result from an appropriate movable singularity at the 'center", but this is not the correct description, and we are basically looking for smooth solutions).

The mathematical (heuristic, numerical, and probably rigorous) resolution of the question of well-posedness of (5)-(7) becomes visible on taking the microcanonical average (volume weighted average, restricted to a flux surface) of the GDE. Without confusion [we have already used a common symbol for $\psi(x,y)$ and $\psi(V)$] we introduce $V(x,y)$ as well as $V(\psi)$, define

$$(8) \qquad \langle\phi\rangle = \oint \phi \frac{ds}{|\nabla V|} = \frac{d}{dV} \int_{V(x,y)<V} \phi \, dx\, dy$$

where

$$\oint \frac{ds}{|\nabla V|} = 1 \quad , \quad \frac{1}{\psi'} = \oint \frac{ds}{|\nabla \psi|} \tag{9}$$

and verify

$$\langle \Delta\psi \rangle = (K\psi')' \tag{10}$$

where

$$K(V) = \langle |\nabla V|^2 \rangle = \oint |\nabla V| ds = \frac{1}{\psi'} \oint (\frac{\partial \psi}{\partial n}) ds \ . \tag{11}$$

Let us detour for a moment with a few properties of the crucial geometrical quantity K(V). First of all, K is fully determined by the family of contours, independent of the distribution of ψ on the contours. Second, for the topology of Fig. 7, K(0) = 0. For a shell $\psi_1 < \psi < \psi_2$, within which $\Delta\psi = 0$, the inductance of the shell is

$$L = \int_{V_1}^{V_2} dV/K \tag{12}$$

It is easily shown that

$$K \geq \ell^2 \geq 4\pi V \tag{13}$$

where ℓ is the length of the contour; the first relation comes from Schwarz' inequality, the second is the isoperimetric inequality (remember, V is area). The first relation is an equality only if $|\nabla V|$ = const. (equidistant curves), the second only for a circle. For a family of ellipses $x^2/a^2 + y^2/b^2$ = const., we have $K = 2\pi V(a/b + b/a)$. For an imminent change in topology at the origin, $\psi = x^2 + \varepsilon y^2 + ay^4 + \ldots$, $K(V) \sim V^{2/3}$ (for $\varepsilon = 0$). Briefly, K is minimized by a family of circles; it is large for elongated ellipses or corrugated contours; it is particularly large for elongated contours in the neighborhood of a degenerate critical point. These qualitative properties of K will allow qualitative estimates of physical behavior under adiabatic compression (and diffusion).

Returning to (10), $(K\psi')' = K\psi'' + K'\psi'$, we note that the center, V = 0, is singular for the averaged Laplace operator since the coefficient of ψ'' vanishes. The fact that no boundary condition can be specified at V = 0 for the 1-D averaged operator, $(K\psi')'$, is a reflection of the fact that V = 0 is an <u>interior</u> point of the domain for the original 2-D elliptic operator, Δ. For the full GDE (6), the coefficient of ψ'' in the averaged equation is K + 1, which does not vanish. The average of the GDE, (5) or (7), is always an ODE (taking K as given). Under appropriate (mild) restrictions, the center, V = 0, is regular for this ODE, allowing $\psi(0) = \psi_1$ to be specified as a boundary condition.

From the above remarks it is heuristically clear why specifying ψ_1 as well as an elliptic boundary condition is likely to be well-posed for the GDE. It is also clear the the <u>averaged GDE</u> must be used in any constructive approximation algorithm to allow the prescribed value of ψ_1 to enter the problem at all. The simplest algorithm (which turns out to converge remarkably well numerically) is to alternate between successive evaluations of the 2-D contours, ψ = const., and the 1-D profile, $\psi(V)$. First, the GDE is used, with right side treated as a given inhomogeneous term (or a given function of V or ψ) in order to determine a family of contours, thence $K(V)$; the values of ψ are discarded. Next the averaged GDE is treated as a second order ODE with $K(V)$ known and is solved (subject to obvious restrictions) taking $\psi = \psi_1$ at $V = 0$ as well as $\psi = 0$ at the outer boundary. The principal result is the solution of a specified equilibrium given adiabatic constraints μ and ν instead of elliptic constraints p and f; as a consequence one also obtains a family of adiabatically related solutions as an external constraint is varied, keeping $\mu(\psi)$ and $\nu(\psi)$ fixed. This can be considered to be the solution of an appropriate (slowly varying) time-dependent problem.

The mathematical significance of the GDE is elucidated by embedding it in a sequence of problems

$$(14)\qquad \begin{cases} \text{(a)} & \Delta\psi = F(\psi) \\ \text{(b)} & \Delta\psi = F(V) \\ \text{(c)} & \Delta\psi = F(\psi') \\ \text{(d)} & \Delta\psi = F(\psi'') \end{cases}$$

In each case we suppose that the domain and the function F are suitably restricted to avoid complications. We now make a number of statements, some proved in restricted formulations, others appear to be amenable to tools of conventional analysis but not yet attempted, still others in the form of supporting nonrigorous analytical and numerical evidence. Case (a) is elliptic. Although (b) is not elliptic (it is not a differential equation), it is qualitatively elliptic in that a solution is determined by an elliptic boundary condition, and the right side is dominated by the left side. The same is true of (c). In other words, it is not the introduction of the highly implicit and nonlinear variable V which is important; it is the appearance of a second derivative V" to interfere with the second order Laplacian, as in (d), which changes the nature of a well-posed problem.

(c) Islation and Generalized Adiabatic Constraints. Now we turn to changes in topology resulting from a variation of constraints. In the elliptic formulation, a change in contour topology is not particularly significant, and the topology of the solution does not enter in existence proofs or practical methods of solution. For example, the elementary equation $\Delta\psi = 1$ will have simple contours in a convex domain, but the topology will change when the domain is sufficiently indented. Similarly, for the nonlinear $\Delta\psi = F(\psi)$, one can have a family of unique solutions in a family of domains with no singular behavior at a change in topology. The variational equation for the elliptic equation $\Delta\psi = F(\psi)$ in terms of

$\phi = \partial\psi/\partial t$ is

$$(\Delta - \dot{F})\phi = 0 \tag{15}$$

This equation makes bifurcation visible, but it exhibits no distinguishing feature at a point of islation. The variational equation for the GDE (6) in terms of $\phi = \partial\psi/\partial t$ is

$$(\Delta + \sigma)\phi = - \langle\phi\rangle'' - \sigma\langle\phi\rangle\ , \quad \sigma = \psi'''/\psi' \tag{16}$$

This equation is singular at a degenerate critical point such as $\psi = x^2 + y^4 + \ldots$ where $\psi'(0) = 0$, or on any curve where $\psi' = 0$ (this is topologically unstable). How to continue the solution past such a point is not obvious from the linear variational equation (which is a GDE, just as is the original nonlinear equation).

The correct treatment of islation (also of anislation - disappearance of an island) is necessarily nonlinear, but nonlinearity alone is insufficient for a change in topology. The matching conditions across a separatrix are, in principle, elementary, but they are subject to a number of subtleties such as differences between $\gamma = 2$ and $\gamma \neq 2$, differences between islands in 2-D and in axial symmetry, and different relationships toward conservation of energy and variational formulations in these various cases.

It is useful to point out that there are apparently two mathematically valid formulations of the equilibrium problem under adiabatic constraints (μ and ν formulation) in a complex geometry. One is to adopt strict flux conservation as suggested by the equations of motion, taking ψ_c as well as ψ_1 and ψ_2 to be given constraints (cf. Fig. 6); $\mu_i(\psi)$ and $\nu_i(\psi)$ are given profiles for $i = 0, 1, 2$, the range of ψ in each function being specified a priori. One can give heuristic iterative and variational arguments to indicate that this strict interpretation of flux conservation will frequently (e.g. in axial symmetry for $\gamma \neq 2$) lead to discontinuities of p, f, and ψ' (ψ is, of course, continuous) across the separatrix. This also implies presence of a surface current, $[\nabla\psi] \neq 0$, at the separatrix. The local pressure balance

$$[p + \tfrac{1}{2} B^2] = 0 \tag{17}$$

will be maintained across the separatrix; this is merely the weak form of (1). Note that although this strict adiabatic formulation may allow one to solve for an equilibrium in a complex topology with strictly given adiabatic constraints μ_i and ν_i, it does not allow the topology to change so long as the profiles $\mu_i(\psi)$, $\nu_i(\psi)$ are strictly preserved.

The surface current at the separatrix is one reason to expect nonconservation of the value of ψ_c (Fig. 6) since the separatrix is perpetually in a resistive boundary layer for any value of the resistivity, no matter how small. On the other hand, Faraday's law implies conservation of ψ_1 and ψ_2 even when they are separated by a resistive layer. Dropping the constraint on ψ_c is the principal ingredient in the generalized adiabatic formulation in which total mass, flux, and volume are preserved, but not in the individual regions, i = 0, 1, 2. The basic matching conditions are found to be

(18) $$[p] = [f] = [\nabla\psi] = 0$$

in addition to (17). Some of the most important auxiliary connection formulas are

(19) $$1/\psi'_o = 1/\psi'_1 + 1/\psi'_2$$

(involving conservation of volume and flux and absence of surface current). In the case of mixing (shrinking domain 0 in Fig. 6a, and growing domain 0 in Fig. 6b), we have

(20) $$\mu_o^{1/\gamma} = \mu_1^{1/\gamma} + \mu_2^{1/\gamma}$$

while for splitting (growing domain 0 in Fig. 6a and shrinking in Fig. 6b)

(21) $$\begin{cases} \mu_1 = (\psi'_o/\psi'_1)^\gamma \mu_o \\ \mu_2 = (\psi'_o/\psi'_2)^\gamma \mu_o \end{cases}$$

The formulas (20) and (21) serve to extend the definition of the "given" function $\mu(\psi)$ when the domain of the independent variable, ψ, changes.

(d) Resistive Diffusion. In the limit of small resistivity, we expect diffusion to be a slow phenomenon compared to wave motion. A formal scaling can be given to decouple the two time scales (actually, two sets of time scales). The principal result is that the plasma is, at every instant, in static equilibrium, $\nabla p = J \times B$ ($\rho du/dt$ is dropped from the equation of motion). However all other time derivatives such as $\partial B/\partial t$, $\partial p/\partial t$ are retained. In other words, the system of equations (which we exhibit later) describes a time dependent (one parameter family) of static equilibria, $\nabla p = J \times B$. Let us examine Ohm's law which takes the form

(22) $$E + u \times B = \eta J$$

where η is the resistivity. At a given instant we can consider B and J to be known, while E and u are not. In an infinite medium ηJ can be arbitrarily split into two parts, E and $u \times B$, by choosing a moving frame. In a bounded domain, the resolution of ηJ into its two components is not arbitrary; this separation is, in fact, the object of the difficult and elaborate theory of diffusion. Consider first the two limiting cases, $\eta J = E$, $\eta J = u \times B$.

If $E = \eta J$, from Maxwell's equations, $\partial B/\partial t + \text{curl } E = 0$ and $J = \text{curl } (B/\mu_o)$, we obtain a simple diffusion equation

(23) $$\frac{\partial B}{\partial t} + \text{curl } (\eta/\mu_o \text{ curl } B) = 0 ,$$

with diffusion coefficient $D_o = \eta/\mu_o$. This equation describes the classical skin effect usually described as field or current penetration into a rigid conductor.

The other extreme, $\eta J = u \times B$ is simplified by assuming that u is perpendicular to B; thus

(24) $$u = -\eta(J \times B)/B^2 = -\eta\nabla p/B^2$$

To further simplify, consider an isothermal case in which p/ρ = const. Conservation of mass, $\partial\rho/\partial t + \text{div}(\rho u) = 0$, therefore takes the form

(25) $$\frac{\partial p}{\partial t} = \text{div}(\frac{p}{B^2} \nabla p)$$

This is a somewhat unusual, nonlinear diffusion equation for p with a (linearized) diffusion coefficient $D_1 = \beta D_o$; (in laboratory plasmas β can take values ranging from 10^{-8} to 10^2). D_1 is frequently termed the "classical" coefficient of plasma diffusion.

One would be tempted to describe (23) as field diffusing through plasma and (25) as plasma diffusing through field. At first glance one might also think that only the relative diffusion matters. But in a domain with boundaries, both plasma and field diffuse (of course, not independently, since they are related by pressure balance). There are thus at least two phenomena, not always distinguishable, and the determination of E and u comes from the solution of a nonstandard global initial-boundary value problem.

The scaling referred to above, in which $\rho du/dt$ was dropped, retains u in other equations of the system. For example, with a simplified treatment of energy conservation, the Grad-Hogan equations are

(26) $$\begin{cases} \nabla p = \text{curl } B \times B/\mu_o \\ \partial B/\partial t + \text{curl } (B \times u) + \text{curl } (\eta/\mu_o \text{ curl } B) = 0 \\ \dfrac{\partial p}{\partial t} + \text{div } (pu) + (\gamma-1)\, p \text{ div } u = 0 \end{cases}$$

We have an isothermal or adiabatic plasma by setting $\gamma = 1$ or $\gamma > 1$ respectively. Evidently (assuming that the system makes sense), u must be determined at each instant in such a way that the time derivatives $\partial B/\partial t$, $\partial J/\partial t$, and $\partial p/\partial t$ are compatible with maintaining $\nabla p = J \times B$. In two dimensions and axial symmetry, $u \cdot \nabla\psi$, the normal component of u, turns out to satisfy a linear GDE (with coefficients depending on the instantaneous equilibrium state). In a sense this reduces the diffusion problem to one that is "known"; but this method of advancing the solution in time is much less efficient, numerically, than others which will be described later.

For purposes of illustration, we turn to a simpler model (which turns out to be the special case of (26) in two dimensions at low β, in which limit p and the two field components decouple). Taking the flux function ψ as in (2), the classical skin effect in a solid conductor is described by the diffusion equation

$$\frac{\partial\psi}{\partial t} = \eta\Delta\psi \tag{27}$$

where η is the (constant) resistivity; the current, $\Delta\psi$, satisfies the same equation. In a deformable plasma, pressure balance requires identical flux and current contours,

$$\Delta\psi = F(\psi,t) \tag{28}$$

The two equation (27) and (28) are, in general, incompatible for any choice of F; (the frequently solved cylindrical diffusion problem in which circular ψ contours automatically coincide with $\Delta\psi$ contours is non-representative). Since a plasma is deformable, we replace (27) by

$$\frac{\partial\psi}{\partial t} + u\cdot\nabla\psi = \eta\Delta\psi \tag{29}$$

and (as a result of scaling or merely for simplicity of the model), we also take

$$\langle u\cdot\nabla\psi\rangle = 0 \tag{30}$$

on each flux contour; (the deformation is incompressible, on the average). It is not immediately evident that the system (28)-(30) with an elliptic boundary condition (say $\psi = 0$ on a fixed or deforming boundary), plus initial specification of $F(\psi,0)$ describes a well-posed problem and determines $\psi(x,y,t)$, $F(\psi,t)$, and $u\cdot\nabla\psi$. (The relation to the GDE formulation is hidden in this model).

A heuristic indication that this dissipative system is well-posed (impractical numerically, but useful for analytic results) goes as follows (cf. [10]). Differentiating the constraint $\Delta\psi = F(\psi,t)$

$$\Delta\frac{\partial\psi}{\partial t} = \frac{dF}{dt} + \dot{F}\frac{\partial\psi}{\partial t} \quad . \tag{31}$$

Taking G to be the inverse (Green's function) of the operator $\Delta - \dot{F}$,

$$\frac{\partial\psi}{\partial t} = G(\frac{dF}{dt}) + g \tag{32}$$

where g is the contribution from the boundary. Averaging on a flux contour

$$\eta F = \langle G\rangle(\frac{dF}{dt}) + \langle g\rangle \tag{33}$$

or

$$\frac{dF}{dt} = \eta\,\langle G\rangle^{-1}F - \langle G\rangle^{-1}\langle g\rangle \tag{34}$$

The Green's function $\langle G\rangle$ is averaged with respect to each pair of variables. For a "reasonable" set of contours, $\langle G\rangle^{-1}$ has been shown to be approximately a Sturm-Liouville operator.* Thus F evolves in time as the solution of a 1-D diffusion-type equation (34). After solving for F, ψ is obtained from (28), and $u\cdot\nabla\psi$ from (29).

The same averaged Green's function technique can be used to formally solve the linear variational equation (16) in the case of the adiabatic GDE

(35) $$\phi = -G(\langle\phi\rangle'' + \sigma\langle\phi\rangle)$$

Averaging and inverting $\langle G\rangle$ yields

(36) $$\langle\phi\rangle'' + \langle G\rangle^{-1}\langle\phi\rangle = -\sigma\langle\phi\rangle$$

The principal part (ignoring bounded operators) is the sum of two Sturm-Liouville operators and is invertible with suitable restrictions.

An alternative solution scheme, which is very efficient numerically (though less tractable analytically), is to average (29), viz.

(37) $$\psi_t = \eta(K\psi')' , \quad (K\psi')' \equiv F$$

If the geometrical coefficient K(V,t) were known, the profile $\psi(V,t)$ would be obtained as the solution of this conventional 1-D diffusion equation. The inductance, K, which reflects the 2-D origin of (37), is recalculated at appropriate time intervals by computing contours from $\Delta\psi = F$ where F is the end result of an interval of 1-D diffusion of (37). Since the geometry usually changes relatively slowly (compared to 1-D profiles which may contain boundary layers, etc.) this formulation successfully solves an unusual, implicit 2-D diffusion equation by the expenditure of not much more computing time than is appropriate for a 1-D standard diffusion problem.

It is instructive to compare (29) with its mean value (37). The convective term $u\cdot\nabla\psi$ is highly implicit and nonintuitive since it arises in response to the changes in shape of the contours required to maintain $\Delta\psi$ constant on ψ contours. The changes in shape are incorporated into (37) (which has no convection) through the inductance coefficient K(V) which, after a little practice, becomes very intuitive.

For example, in Fig. 8 are presented the results of a numerical solution of (28)-(30), using the methods just indicated. The initial state, Fig. 8a, represents a belt pinch with J(V) = const. [the diagram on the right is J(V)]. A pair of coils in the vacuum region between the plasma and the rectangular box is used to compress the waist of the plasma. Figure 8b is just before islation and Fig. 8c some time later. The corrugation near the edge of the plasma increases the value of K and gives a small current increase toward the plasma edge (the boundary condition has been chosen to eliminate the skin current which would give a sharp

* Private communication, G. Vigfusson.

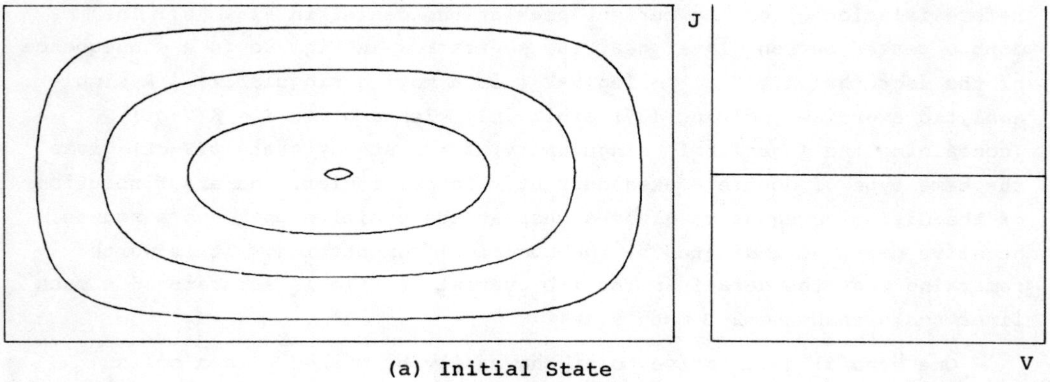

(a) Initial State

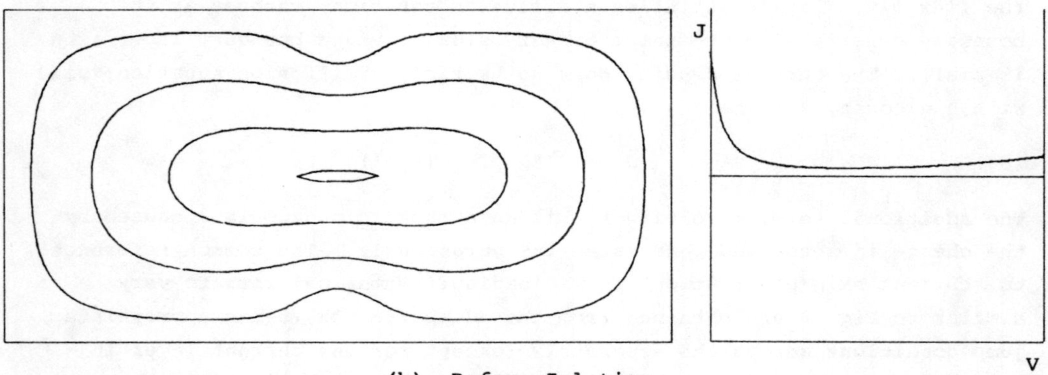

(b) Before Islation

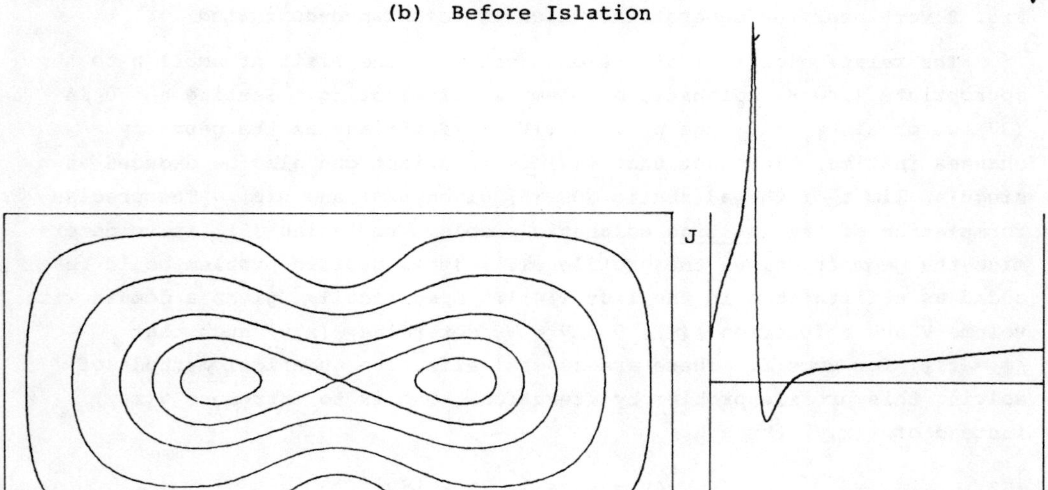

(c) Fully Developed Islands

Figure 8: Transition from Belt Pinch to Doublet.

negative peak). The elongation of the flux contours near the center just before islation gives the current peak at the center in Fig. 8b. The double peaked current layer near the separatrix in Fig. 8c is a consequence of the fact that $K(V-V_c) \sim -\log|V-V_c|$ is a moving singularity. A simple analytic exercise, solving (37) explicitly with a given $K = K(V-ct)$ (containing the logarithmic singularity) for a steady state $\psi(V-ct)$ gives the same type of double peaked current. In particular, the exact solution of the diffusion equation gives a cusp at the positive peak and a rounded negative peak, as indicated by the numerical computation. It is worth remarking that the detail in the 1-D current profile is accurate on a much finer scale than the 2-D mesh size.

One crucial qualitative point can easily be missed. In a solid conductor, flux and current diffuse identically. In a deformable plasma, the flux $\psi(V,t)$ again satisfies a diffusion equation; changes at the boundary penetrate to the interior very slowly, via a boundary layer if η is small. The current density does <u>not</u> satisfy a diffusion equation; viz. with η = const, $J \equiv (K\psi')'$,

$$J_t = \eta(KJ')' + (K_t\psi')' \tag{38}$$

The additional term, involving K_t, is an <u>adiabatic</u> change in J caused by the change in shape and penetrates instantaneously. The common reference to "current skin penetration" is misleading. Numerical results very similar to Fig. 8 are obtained from the adiabatic GDE, using appropriate jump conditions across the separatrix (except for the current layer in Fig. 8 very near the separatrix, which is resistance-dominated).

The relation of this diffusion problem in the limit of small η to the appropriate (low β) adiabatic problem is illuminating. Setting $\eta = 0$ in (37) we obtain $\psi_t = 0$; the profile $\psi(V)$ is invariant as the geometry changes in time. The fact that $\psi(V)$ is invariant can also be deduced as a singular limit of the adiabatic GDE (5) given $\mu(\psi)$ and $\nu(\psi)$. The precise formulation of the <u>limiting</u> adiabatic problem [replacing (5)] is to determine the geometry given the profile $\psi(V)$; [this profile problem could be added as a fifth item in the list (14)]. Specifically, given a domain with volume $\bar{V}$ and a function $\psi(V)$, $0 < V < \bar{V}$, determine $\psi(x,y)$ such that $\Delta\psi = f(\psi)$ for some f. There are several effective numerical methods of solving this profile problem by iteration. One is to introduce $V(x,y)$ instead of $\psi(x,y)$ through

$$\Delta V = K' + \frac{\psi''}{\psi'}(K - |\nabla V|^2) \tag{39}$$

Here $\phi(V) = \psi''/\psi'$ is given and $V(x,y)$ is to be found. More interesting is the following variational formulation. Take the conventional variational function (Dirichlet Principle)

$$U = \int \frac{1}{2}|\nabla\psi|^2 \, dx\, dy \tag{40}$$

where admissible functions $\psi(x,y)$, defined in a simple 2-D domain, are assumed to be compatible with a given $\psi(V)$. With this class of admissible functions $\psi(x,y)$, U turns out to be stationary for $\psi = \psi_o(x,y)$ such that ψ_o = const. and $\Delta\psi_o$ = const. contours coincide. A member of the admissibility class can be alternatively described by specifying a family of curves which simply cover the domain as in Fig. 7; given the curves one calculates V and assigns ψ according to $\psi(V)$, then calculates $\psi(x,y)$, hence U. In this formulation the geometry and profile parts of the various formulations (14) are visibly distinct.

As a final comparison between the resistive and adiabatic problems we present a table of singular behavior near a separatrix:

Adiabatic	Resistive
$J \sim V^{-1/3}$ (acute)	$J \sim 1$
$K \sim 1$	$K \sim \log \lvert V \rvert$
$K' \sim V^{-1/3}$ (acute)	$K' \sim 1/V$
$\psi' \sim 1$	$\psi' \sim 0$

The statement $J \sim V^{-1/3}$ (acute) means that J has this unbounded behavior on the acute side of the separatrix, but is bounded on the obtuse side. Note that $J \sim 1$ near an acute corner automatically implies that $V \sim \psi \log \psi$; to have $\psi' \sim 1$ [for the adiabatic problem to be meaningful, $p = \mu(\psi')^\gamma$ cannot be zero] J must be unbounded. J is more singular in the adiabatic case, K in the resistive. There is evidently a very complicated transfer of singularities in the limit $\eta \to 0$ [e.g. for very small η there is a transfer from $K' \sim V^{-1/3}$ (acute) near the separatrix to $K' \sim 1/V$ very near the separatrix).

One physical conclusion is worth repeating. The time scale for magnetic field line "breaking" and "reconnection" (which are terms used in astrophysics) is not related to that of any dissipative mechanism; compared to the latter, changes in topology occur arbitrarily rapidly. This has been clearly demonstrated, analytically and by adiabatic and resistive numerical calculations.

References: For a compact survey of the GDE and adiabatic theory see [18] and similarly for diffusion [14] and [20]. The earliest reference to plasma induced (rather than magnetic error) islation was in 1967 [6], the first heuristic mathematical GDE theory (applied to diffusion) was in 1970 [10], following the introduction of the slow time scale diffusion formulation (without inertia) in [9]. The connection with the continuous spectrum was mentioned in 1971 [12], in more detail in [15]. In the more physical literature concerning diffusion, see [2], [5], [21].

CONCLUSION. In a period of over twenty years, a new scientific discipline has arisen, based on the goal of magnetically confining a hot plasma under conditions appropriate for the release of energy by nuclear fusion. New experimental techniques have been developed and refined by orders of magnitude; new technologies have begun; new (as well as applications of old) physical and mathematical theories have appeared and in some cases reached a level of scientific maturity. Effective use is being made of large scale computing for instantaneous and retrospective interpretation of the overwhelming quantity of experimental data; computer use as an imaginative supplement to theoretical analysis and as an independent source of physical and mathematical "experimental" information through mathematical modeling have already had significant impact on the understanding of plasma phenomena.

The almost incredible advances in the accuracy of physical measurements and in control of plasma, and to a lesser extent in externally visible plasma confinement landmarks and in plasma understanding are the most obvious indications that we shall successfully overcome the (still undefined) enormous difficulties that are certainly ahead. Equally important is the flexibility inherent in the almost unlimited variety of plausible options that are available; wherever one obstacle has appeared, three solutions have presented themselves. The present state-of-the-art is sufficiently fluid so that maximal orderly speed towards the goal of economical fusion power requires examination of as many carefully selected options as financial exigencies will permit, rather than arbitrarily closing down on the one or two which appear to be "most likely". In any event, there is no doubt that the ultimate future, whatever the date, lies in fusion taking the position as prime energy resource from oil and coal, with fission as a necessary but unpleasant holding action, and solar, geothermal, etc. as supplemental resources, depending on local conditions.

Mathematically, the welding of numerical experimentation with traditional analysis to extend the power of mathematics in order to be able to act effectively on a practical time scale is only beginning to make itself felt. Since the digital computer is essentially a purely mathematical device (more precise and demanding of rigor than many mathematicians), this relationship is a natural one. Perhaps the most significant feature, practically, is that no conceivable future generation of computing machines is large enough to allow routine solution of plasma reactor problems; however, optimization, efficiency, and cleverness at the analysis-computing interface empirically have been found to be worth many orders of magnitude (or generations of computer hardware).* In particular, to determine what is far in the future or may be soon at hand is not a simple evaluation.

---

* In another context, see Ref. [22].

REFERENCES

[1] A. S. Bishop, Project Sherwood - The U.S. Program in Contolled Fusion, Addison-Wesley, Massachusetts, 1958.

[2] D. Pfirsch and A. Schlüter, Influence of electrical conductivity on the equilibrium of low pressure plasmas in Stellarators, Max-Planck-Institüt für Physik und Astrophysik, MPI/PA/7/62, München, 1968. Unpublished.

[3] H. Grad, Mathematical problem in magneto-fluid dynamics and plasma physics, Proceedings of the Intl. Conference of Mathematicians, Stockholm, 1962, 560-583.

[4] C. K. Chu and H. Grad, Magnetofluid dynamics, in Research Frontiers in Fluid Dynamics, eds., R. J. Seeger and G. Temple, John Wiley and Sons, New York , 1965, 308-349.

[5] H. P. Furth, J. Killeen, and M. N. Rosenbluth, Finite-resistivity instabilities of a sheet pinch, Phys. Fl., 6 (1963) 459-484.

[6] H. Grad, Toroidal containment of a plasma, Phys. Fl., 10 (1967) 137-154.

[7] H. Grad, Plasmas, Phys. Today, 22 (1969) 34-44.

[8] H. Grad, Problems in magnetostatic equilibrium, Proceedings of the Intl. Colloquium on Magnetohydrodynamics, Lille, 1969; also report MF-62, Courant Institute of Mathematical Sciences, April, 1970.

[9] H. Grad and J. Hogan, Classical diffusion in a Tokamak, Phys. Rev. Letters, 24 (1970) 1337-1340.

[10] H. Grad, Mathematical problems arising in plasma physics, Proceedings of the Intl. Congress of Mathematicians, Nice, 1970, Vol. 3, 105-113.

[11] W. C. Gough and B. J. Eastlund, The prospects of fusion power, Sci. Amer., 224 (1971) 50-64.

[12] H. Grad, Plasma containment in closed line systems, IAEA, Vienna, 1971, 229-239. IAEA-CN-28/J-7.

[13] H. Grad, Magnetic properties of a contained plasma, Transactions, N. Y. Acad. of Sci., 33 (1971) 163-177.

[14] H. Grad, Classical plasma diffusion, in Advances in Plasma Physics, Vol. 5, John Wiley and Sons, New York, 1974, 103-119. [Original Russian publication, 1971.]

[15] H. Grad, Magnetofluid-dynamic spectrum and low shear stability, Proceedings of the Natl. Acad. Sci., 70 (1973) 3277-3281.

[16] F. L. Ribe, Fusion reactor systems, Rev. Mod. Phys., 47 (1975) 7-41.

[17] W. R. Ellis, Reactor scaling of linear theta pinches, Proceedings of the High Beta Workshop, ERDA 76/108, New Mexico, 1975, 365-389.

[18] H. Grad, P. N. Hu, and D. C. Stevens, Adiabatic evolution of plasma equilibrium, Proceedings of the Natl. Acad. Sci., 72 (1975) 3789-3793.

[19] Fusion Power by Magnetic Confinement Program Plans, ERDA-76/110/1, July, 1976.

[20] H. Grad, P. N. Hu, D. C. Stevens, and E. Turkel, Classical plasma diffusion, Proceedings of the 6th Intl. Conference on Plasma Physics and Controlled Nuclear Fusion Research, Berchtesgaden, 1976, IAEA-CN-35/B9, to appear.

[21] R. B. White, D. A. Monticello, M. N. Rosenbluth, and B. V. Waddell, ibid., IAEA-CN-35/B2-1, to appear.

[22] D. E. Knuth, Mathematics and computer science: coping with finiteness, Science, 194 (1976) 1235-1242.

THE COURANT INSTITUTE OF MATHEMATICAL SCIENCES
NEW YORK UNIVERSITY
NEW YORK, NEW YORK 10012

Proceedings of Symposia in Applied Mathematics
Volume 21
1977

NUCLEAR ENERGY - PROBLEMS AND PROMISE

MILTON S. PLESSET

A presentation to members of the American Mathematical Society should include areas to which mathematics can make contributions. In the field of nuclear energy we can point to the very important contributions which have been made by applied mathematicians to transport theory, in particular to neutron transport theory. While many important and interesting problems remain in that field, I believe that from a practical point of view the engineer feels that the techniques that are now available to him are quite sufficient for his design needs. There are, however, problems that remain for which we do not have an adequate basis as yet for quantitative solution. I am referring now to those two-phase flow problems which arise in the study of "off-design" behavior of a nuclear power plant. You have heard of the studies of Emergency Core Cooling Systems in connection with nuclear reactor plants. These studies bear on LOCA (Loss of Coolant Accident) and the subsequent performance of ECCS (Emergency Core Coolant Systems). For further information on this subject I refer the reader to the well-known "Rasmussen Report" which is available as an extensive publication, WASH-1400 (1). A commentary on the preliminary, or draft, form of this report has been prepared under the auspices of the American Physical Society (2). In LOCA, one encounters two-phase flows, liquid and vapor, for which one needs both the hydraulic and the heat-transfer properties. One difficulty with a quantitative determination of these properties may be made evident by pointing out that these two-phase flows can exist in quite different topological states which I may list here to illustrate the point:

1. Slug flow - each phase of macroscopic extent well-separated and distinct from regions of the second phase;
2. small bubbles of vapor or gas in the liquid;
3. droplets of liquid in the vapor or gas.

It is of course evident that the exchange of mass, momentum, and energy between the phases will be very different from the different possible flow configurations, and these differences are important for the analysis of the flow and the heat transfer in the loss of coolant accident and the subsequent re-flood of the nuclear core from emergency sources of liquid coolant. In the present "state of the art" the analysis is replaced by so-called models. In view of the lack of an explicit physical basis for many features of these models, many elements of conservatism are imposed on the design and operation of nuclear reactors.

I have discussed the problem of two-phase flow here as an example of a subject that is in need of development, but I do not want to give the impression that our lack of knowledge reflects a lack of contributions from the mathematician. Before applied mathematics can contribute to our needs here, we must have some additional experimental information so that the field may be treated in an orderly and consistent manner. This information must be furnished by the physicist and the engineer so that there is an adequate basis for mathematical development.

Since I cannot present you with new problems in a way suitable for your consideration, I propose to present instead some information which may be useful to you in participating in decisions regarding nuclear energy which are, I believe, of great societal importance. The promise of nuclear energy is well-known and arises from the abundance of a resource, primarily uranium and secondarily thorium, which has no other significant use. The basic material for the production of nuclear energy, therefore, differs in this respect from coal or oil which are valuable for applications other than for the generation of electrical energy. Nuclear energy was initially viewed not only as an abundant energy resource but also as one which would be very cheap. While nuclear energy is still significantly cheaper than electrical energy produced from combustion of fossil fuels, the economic margin is less than originally expected. Figure 1 indicates a comparison of costs. The reason that nuclear energy costs are higher than first expected is that the construction expense has increased, in no small degree because of the expense of redundant safety measures which are now required before a license to operate is given. Figure 2 gives some idea of the size and complexity of a commercial nuclear power plant. The barriers between the environment and the nuclear fuel with its associated radioactivity consist first of the zircaloy clad which encloses the fuel elements, second of a steel pressure vessel which encloses the entire fuel core, and finally of the concrete containment structure which is lined with steel. The capital cost of such a plant which would have an electric generating capacity of 1,000 megawatts would be about one billion dollars. The strictly nuclear components would incidentally represent only 10 per cent of this capital cost. Heretofore, coal burning plants have been considered to require an appreciably smaller capital investment; it is now becoming apparent that the capital cost of a coal burning plant is no less than that of a nuclear plant (Fig. 1 to the contrary). The reason for this rise in cost is the realization that the emission of particulates in these plants is a health hazard which can no longer be tolerated. The precipitators and scrubbers, which were thought to be effective for control of particulate emissions, do not eliminate the extremely small particles which produce serious lung damage in man. Additional and rather expensive treatment of stack discharge is necessary.

The installation of nuclear power plants has been strongly opposed by some individuals and groups for a variety of reasons. I do not propose to discuss in detail one argument which is put forward against nuclear power which says that additional electric power in unnecessary. It is argued that power needs can be greatly reduced by improved energy conservation and by the introduction of alternate energy production methods such as fusion, solar, or geothermal. One cannot have the slightest reservation about additional efforts directed toward energy conservation. These discussions of energy conservation further emphasize the high overall energy consumption in the United States. This point is correct, but ignores the fact that the United States does not have the leading position among world countries in installed electric power capacity per person. In this connection I present a correlation in Fig. 3 of installed electric power per capita with life expectancy for several countries. Another correlation of electric power consumption with employment is shown in Fig. 4. Let me present these correlations in the modest spirit that should accompany most'such correlations.

With regard to alternative forms of generation of electric energy, however, they are subject to very important uncertainties. Fusion energy has not yet been shown to be feasible. If it is shown to be practicable, the economics must then be analyzed. In any case, it must be put well in the future. Solar energy can contribute to some of our energy needs, but for large scale electric power production it is limited by cost and by environmental problems. Similar limitations also apply to geothermal energy.

To return to the consideration of problems of consequence with nuclear power production, there are two which generally appear paramount in the mind of the general public — the problem of waste disposal and the problem of safety, and I propose to consider these briefly.

For the problem of the radioactive waste products which accompany nuclear power production, considerable emphasis has been put on plutonium disposal. Any discussion of plutonium is inevitably charged with considerable emotion, perhaps because plutonium is associated with nuclear bombs. A second feature that distinguishes plutonium is its half-life of 25, 000 years. Since one may use a rough rule of thumb of 10 half-lives as a time to an innocuous level of activity, it would appear that plutonium must be sequestered for times of the order of 250, 000 years or more. About 250 kg of plutonium is produced in commercial nuclear power plants per year and this plutonium is, to a large extent, separated from other radioactive wastes. The separation is in anticipation of the possibility that it may be recycled into nuclear power plants and consumed. Such recycle would decrease the need for uranium fuel. It must be pointed out, however, that a complete removal of plutonium from waste is not feasible and one must anticipate that of the order of 0.1 per cent of the plutonium will remain with the other waste products. Further, if plutonium recycle is not approved, then all the plutonium produced will ultimately need disposition.

It has been widely stated that "plutonium is the most poisonous substance known to man" and that "there is enough plutonium produced in a nuclear power plant in a year to give a fatal dose to every living person". The answer to this latter remark that is sometimes given is that "there is enough air in a small room to kill every living person". Both such statements are meaningless. It is true that a small air bubble in a person's circulatory system can be fatal just as it is true that a suitable implantation of a small amount of plutonium, or of a variety of other substances, can lead to death. The toxicity to man of plutonium has been studied for some time (4, 5, 6), and it must be stated at the outset that there is no known chemical toxicity of plutonium. The hazard, therefore arises only from its potential for causing cancer by the somatic damage from its alpha particle emission. Plutonium does not cause any genetic damage. The carcinogenic damage can arise from deposition in the lung by inhalation or from deposition on the bones. The permissible concentrations of plutonium are similar to those for thorium or for radium, traces of which occur in all diets and all drinking water. The permissible body burden for these materials is established by the International Committee for Radiological Protection and represents the amount of a radioactive substance which can be deposited in the body without significantly jeopardizing health. The permissible body burden for radium, for example, is 100 nanocuries, and the permissible body burden for plutonium is 40 nanocuries (7). The difference between these doses of plutonium and radium comes from the tendency of plutonium to deposit at the bone surface while radium tends to distribute in the interior. It might be pointed out that plutonium, unlike radium, is not absorbed in the digestive tract, and it is not absorbed through the skin. The principle hazard is through inhalation as small particles, and deposition in the lung. The residence time in the lung is a few years where the carcinogenic risk is similar to other alpha particle emitters (4, 5, 6). It may also be pointed out that plutonium does not tend to be taken up by plants or animals. The risk of plutonium reduces to possible inhalation as small particles for which vaporization and dispersion would be prerequisite. Present means of handling and use of plutonium cannot lead to such vaporization or dispersion. Plutonium is a solid with very low vapor pressure so that vaporization and dispersion would require an intense explosion such as occurs in a nuclear explosion. A nuclear explosion in a commercial power plant is physically impossible. An extreme malfunction of the plant could lead to melting of the core, but it is highly improbable that any plutonium would go beyond the confines of the containment structure (1).

We should not minimize the need for careful handling and disposition of the radioactive wastes from nuclear power plant operation, but we should put our emphasis on the significant aspects of the problem. The radioactive constituents of the waste which must be considered foremost are strontium 90 and cesium 137. Both of these radionuclides are present in significant quantitites in the waste, and both enter the biosphere since they are taken up by plants and animals. Further, their half-lives are about 30 years. Of these two, strontium is the more serious since upon deposition in bone it has a long residence time in the body while the residence time for cesium is appreciably shorter. Clearly we should require that these substances be kept out of the biosphere for at least 300 years.

In the public discussion of the waste and waste-storage problem there has been concern expressed that a "solution has not yet been found." Actually the problem is being approached in a considerate and deliberate manner. The requirements for acceptable storage are being formulated so that suitable disposal methods can be accurately determined. For a brief exposition of the necessary properties of a long-term repository, reference may be made to the work of B. Cohen (8). The deliberate approach to the disposal problem is appropriate since the mass of material which requires storage is small and can be easily handled in surface storage for some time. In thinking about a waste problem we are too accustomed to thinking in terms of the waste discharges from a coal burning plant, or from a large chemical industry. A coal burning plant with a capacity of 1,000 megawatts (electric) will consume 2.5 million tons of coal per year and will generate hundreds of millions of pounds of oxides of sulfur and nitrogen as well as similar quantities of ash. If the oxides of sulfur are not to be emitted to the atmosphere, limestone scrubbing will produce of itself hundreds of thousands of tons of sludge per year. A nuclear power plant, by contrast, will produce only a few tons of waste per year so that the waste volume is less by a factor of the order of a million. We do not mean that the nuclear waste is to be taken lightly; like other toxic wastes from industrial activity it must be carefully isolated from the environment. One may say, however, with some assurance that the waste disposal requirements for nuclear power systems will be met at moderate cost. In view of the small volume of nuclear waste, it could be kept for some time in secure surface repositories, and this procedure is now being followed. For the long term the waste material would be deposited underground in solidified form. The solid matrix gives very low leaching rates into the surroundings which will be well-isolated from sources of ground water (8). The essential requirement for an underground repository is that the waste should be isolated until its radioactivity falls to the natural level in the earth, or below. Promising geologic formations are deep shale rocks or some deep salt deposits.

The second major area of public concern is with the safety of nuclear power plants. It must be said at the outset that the safety record for the operation of these plants has not been equalled in any other major industrial activity. There are over 50 commercial nuclear power plants operating in this country and about the same total number operating in Western Europe, the United Kingdom, and Japan. These nuclear reactors combine to give about 1,000 reactor-years of operation in which no member of the public has been exposed to any harm or injury. The public has, however, been told of hypothetical accidents which could cause harm to persons or property. Representatives of various groups therefore have asked that nuclear power plants be certified as "perfectly safe". Such complete assurance, of course, cannot be given.

There are several aspects of the safety question for any significant societal activity. The first, of course, is the frequency of occurrence of an accident together with the associated cost in harm to people, to the environment, or to property. The general public often becomes too familiar with the risks associated with many conventional activities and ignores them. In this acceptance of the familiar risks, the public, unconsciously in many cases, weighs them against more or less obvious benefits. A familiar example of an accepted risk is the development of a community below a dam. It is presumed that the dam represents a significant benefit, but dam failures, like earthquakes, occur with nearly predictable frequency and are often accompanied with considerable loss of life and destruction of property.

The possible consequences of a major accident with a nuclear power plant have been given vivid descriptions, but these descriptions are not accompanied by discussions of probabilities nor by presentations of benefits to society. The probability of a major accident with a nuclear power plant is known to be extremely small, and considerable effort has been made to evaluate this probability. The most extensive study available to us is given in the Rasmussen Report (1) and some results from this study are shown in Fig. 5. At the same time an estimate of the benefit awareness of the public has been made (9), and compared with the benefit awareness of other risks (see Fig. 6). It is clear from the experience to date and from all the serious efforts to analyze possible malfunctions the probability of a release of damaging radioactivity is extremely small. At the same time the benefit of an abundant supply of electrical energy is not perceived by many persons in our country. I do not refer here to the somewhat luxurious domestic consumption by the middle class and the wealthy, but to essential needs of commerce and industry. Our society has a greater dependence on this energy than is apparent to many individuals. To reduce this supply or to increase its cost could impose serious strains on the structure of our society.

I hope that you will view Figs. 5 and 6 as attempts to give a quantitative flavor to ideas which are essentially qualitative in nature. If I may be allowed to continue with qualitative ideas, I would like to conclude with remarks on a well-known public health problem which actually was the initial activity of the Rockefeller Foundation health program. This program undertook to attack a disease that was caused by the hookworm parasite. The disease appeared to be an ideal example for eradication by a public health program. The life history of the hookworm as well as its means of entry into man was known. A simple, rapid cure was developed which cost only a few cents to administer. The disease was attacked vigorously in the early part of this century in the southern portion of the United States where it was widespread. In spite of the knowledge of the disease and the inexpensive, rapid cure, it stubbornly resisted eradication. It disappeared, however, quite abruptly in the South after the installation of TVA electric power complex (10). Following the distribution of electricity into an area previously without it, water pumps and flush toilets became common, as did shoes for the poor people. Hookwork disease then became a thing of the past as something better forgotten than remembered.

REFERENCES

1. Reactor Safety Study - An assessment of Accident Risk in U. S. Commerical Nuclear Power Plants, Report WASH-1400, National Technical Information Service, Dept. of Commerce, Springfield, Va. 22151, (1974).

2. Report to the APS by the Study Group on Light-Water Reactor Safety, Reviews of Modern Physics, Vol. 47, supplement No. 1, (1975).

3. Starr, C., A Strategy for National Electricity Production, Electric Power Research Institute, Palo Alto, CA., (July 10, 1975).

4. The Toxicity of Plutonium, Medical Research Council, London, Her Majesty's Stationery Office, (1975).

5. Bair, W. J. and Thompson, R. C., Plutonium: Biomedical Research, Science, Vol. 183, (1974), pp. 715-722.

6. Bair, W. J., Richmond, C. R., and Wachholz, B. W., A Radiobiological Assessment of the Spatial Distribution of Radiation Dose from Inhaled Plutonium, WASH-1320, U. S. Atomic Energy Commission, U. S. Government Printing Office, (1974).

7. For the United States, the permissible limits have been set at levels below that imposed by the International Committee for Radiological Protection; see the U. S. Code of Federal Regulations Title 10, part 20. In particular, these limits are translated into limits on ground water contamination.

8. Cohen, B., Physics Today, January (1976)., pp. 9-14.

9. Starr, C., Redman, R. and Whipple, C., Philosophical Basis for Risk Analysis, Electric Power Research Institute, Palo Alto, CA., (1975).

10. Williams, G., The Plague Killers, Charles Scribner's Sons, (1969).

Professor Milton S. Plesset
Department of Engineering Science
California Institute of Technology
Pasadena, California 91125

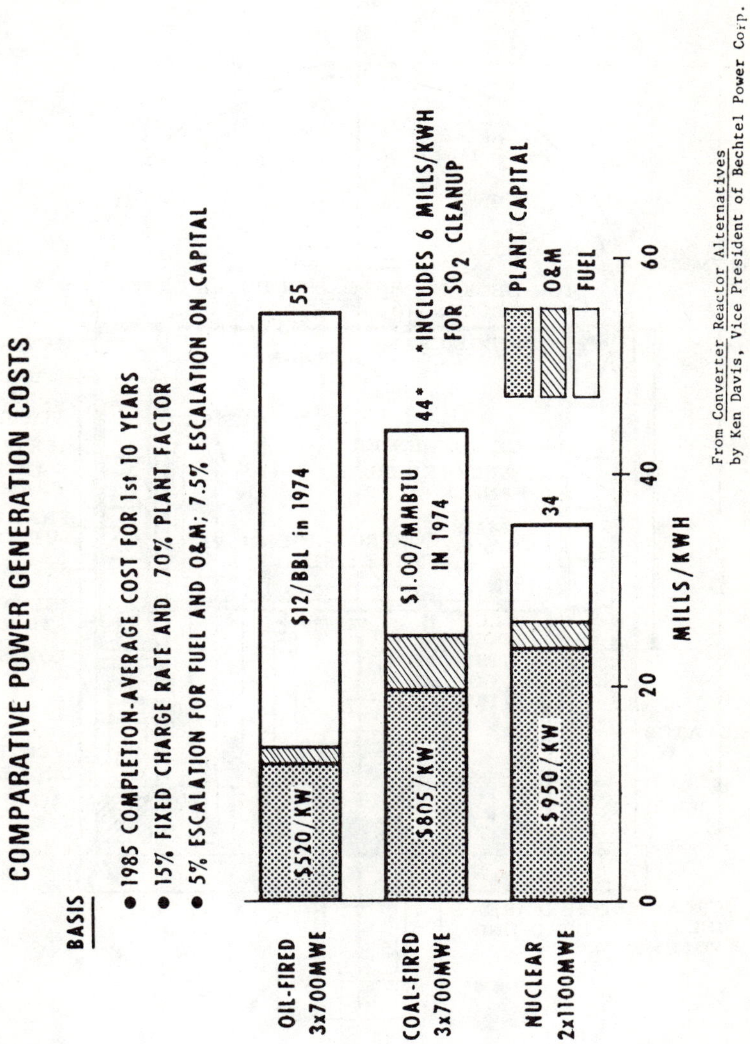

Fig. 1. Comparative Power Generation Costs

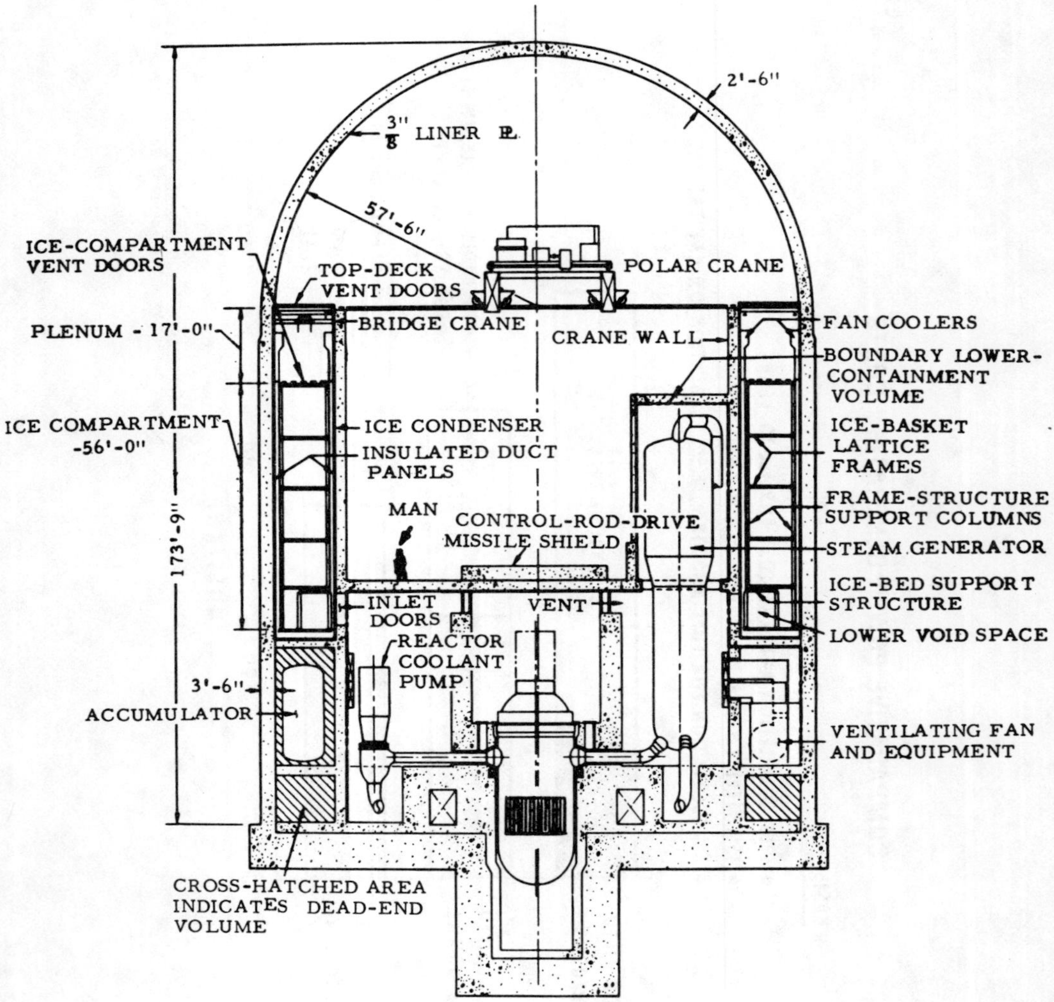

Fig. 2. Typical Primary Coolant System and Containment for a Large PWR System.

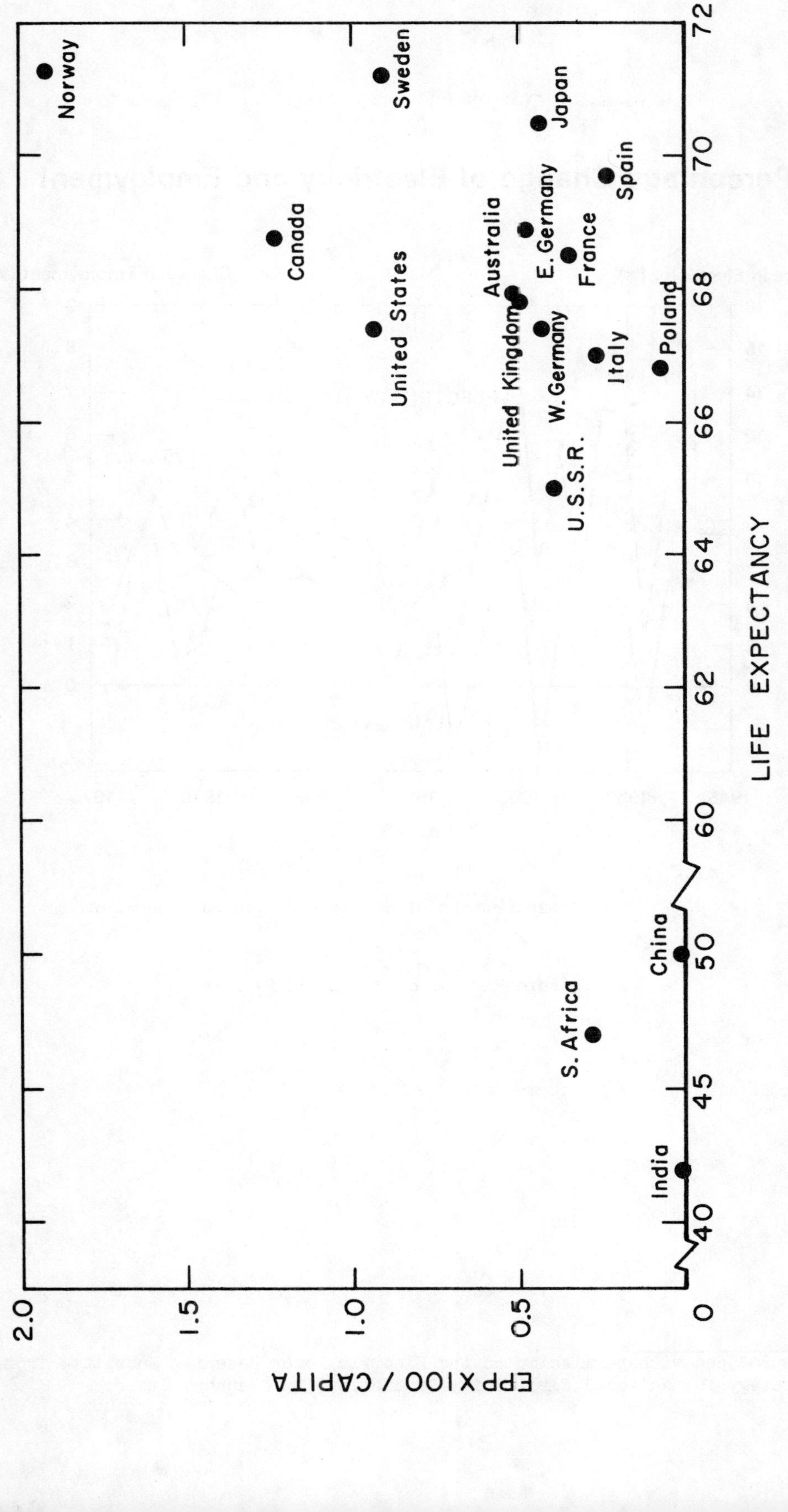

Fig. 3. Life Expectancy

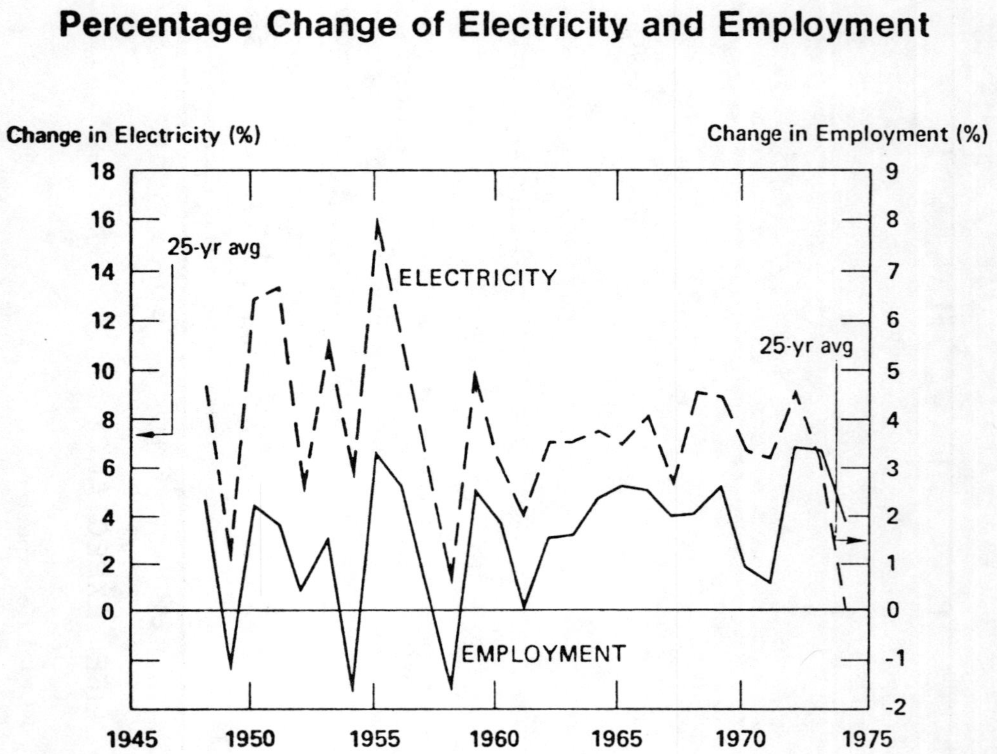

Fig. 4. Percentage Change of Electricity and Employment.*

Taken from Reference 3 - EPRI Report

* Reproduced with permission of the Electric Power Research Institute from "Strategy for National Electricity Production" by Chauncey Starr.

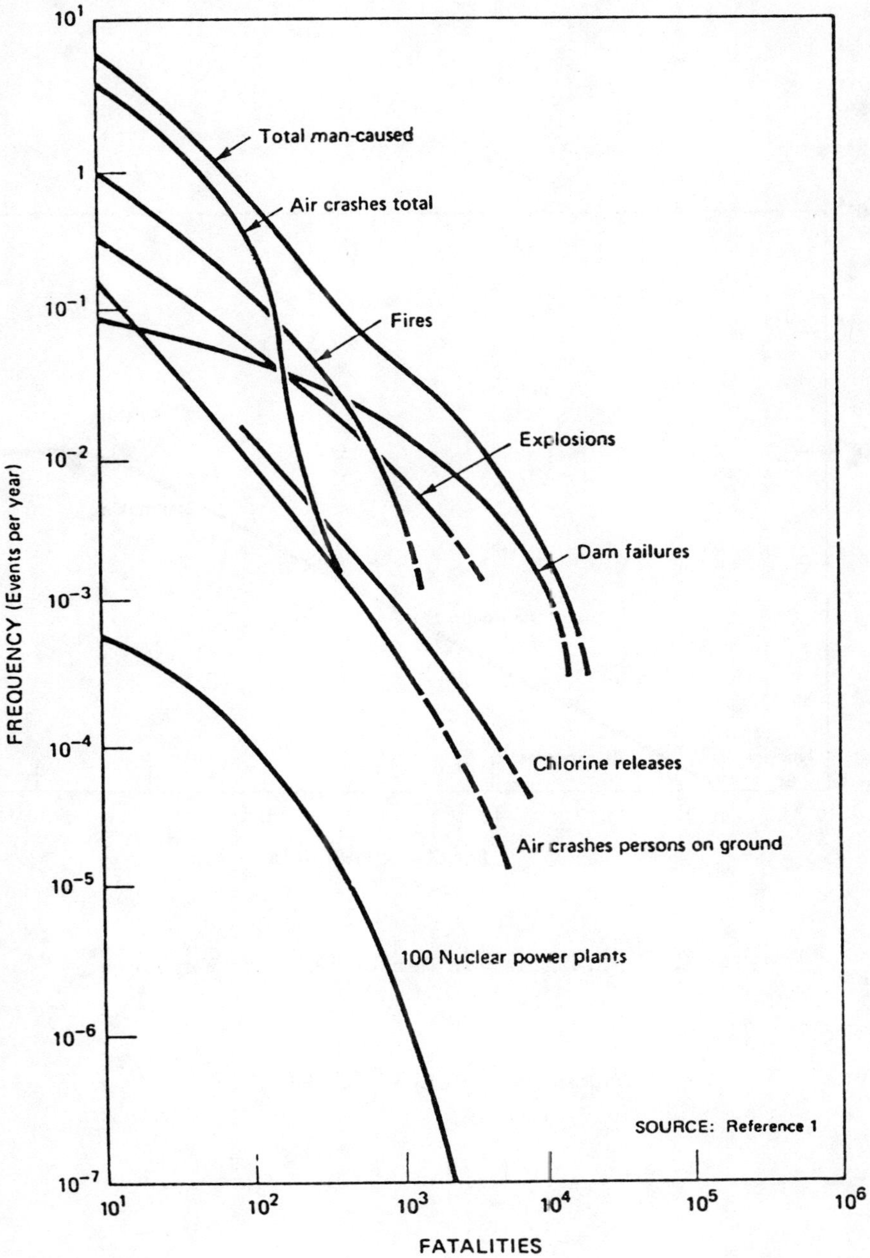

Fig. 5. Frequency (Events per year)

Taken from Reference 6.

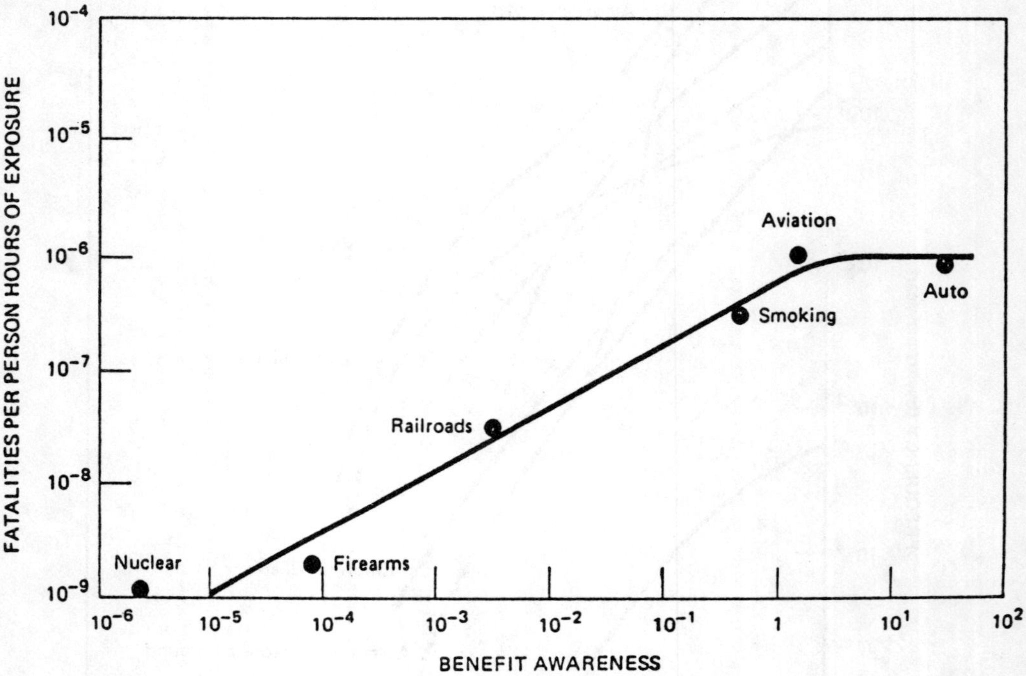

Fig. 6. Fatalities per Person Hours of Exposure*

Taken from Ref. 9 - EPRI

* Reproduced with permission of the Electric Power Research Institute from "Philosophical Basis for Risk Analysis" by Chauncey Starr, Richard Redman, and Christopher Whipple (1975).

Proceedings of Symposia in Applied Mathematics
Volume 21
1977

LASER FUSION

F. D. Tappert

ABSTRACT

Laser fusion is a scientific and engineering endeavor to obtain useful amounts of energy from thermonuclear fusion reactions taking place in inertially confined microscopic fuel pellets which are heated and compressed by high power laser radiation. Among the problem areas where methods of theoretical physics and mathematical analysis are being successfully applied are the following: laser theory and quantum optics; nonlinear optical propagation and diffraction theory; and hydrodynamic and radiative transport theory. In each of these areas, the physical behavior is described by complex systems of nonlinear partial differential equations. Considerable physical insight has been obtained by asymptotic analyses and explicit solutions of these equations in special limiting cases. A more powerful and general approach which has been developed by the laser fusion research laboratories is to construct large computer codes which take into account the many complicated interacting physical processes. The design of such computer codes requires efficient algorithms for the numerical solution of partial differential equations, optimal use of high-speed computing machine architecture, and effective computer graphics facilities to display the large amount of information produced by these codes.

* Research partially supported by the U. S. Energy Research and Development Administration, Contract No. EY-76-C-02-3077.*000.

INTRODUCTION

The technological development in the last twenty years of high energy coherent sources of electromagnetic radiation in the visual part of the spectrum has made profound changes in almost every aspect of experimental and practical science. The most exciting of these applications of laser technology is in the generation of energy from nuclear fusion reactions. In conception, a laser-induced fusion plant would operate much like an internal combustion engine: the small pellet of thermonuclear fuel is suspended in the reaction chamber (the "piston") and ignited by a convergent set of laser beams (the "spark"). This process is repeated with a number of pistons which simultaneously convert the heat produced in the thermonuclear fusion micro-explosions into electricity.

The important difference between laser fusion and conventional energy sources is the tremendous energy density that is required for ignition to occur, and thus tremendous concentrations of energy are obtainable from fusion reactions. The energy density of the explosion in an automobile piston is about *one ten-billionth* of that required in a laser induced fusion reaction, and a tankful of deuterium (eighteen gallons, say) could supply the energy necessary for all of the electrical energy used in the whole of the United States for a month. This difference in energy density results also in a new realm of physical processes which become important -- processes whose theoretical description has confronted physicists and mathematicians with the problem of nonlinear systems more forcefully than any problem in the past.

The theoretical description of the laser fusion process can be usefully divided into two aspects which have distinct characteristics: the generation and propagation of the laser pulse (the "ignition" system); and the plasma dynamical processes occurring within the fuel pellet (the "combustion" system). In both of these areas, the physical behavior is described by complex systems of nonlinear partial differential equations. Considerable physical insight has been obtained by asymptotic analyses and explicit solutions of these equations in special limiting cases. A more powerful and general approach which has been developed by the laser fusion research laboratories is to construct large computer codes which take into account the many complicated interacting physical processes.

Easily accessible semi-popular accounts of the basic physical principles involved in laser fusion research can be found in Refs. [1]-[4]. A good technical review article is Ref. [5]. A long-range government planning document which discusses the ERDA funded laser fusion programs is available [6]. The current status of laser fusion research at the two leading U. S. laboratories can be found in their 1975 annual reports [7]-[8].

LASERS

The laser itself consists of basically two components, an active medium which emits the radiation at a characteristic frequency and an optical resonator which builds the energy from these emissions into a coherent, collimated beam of light. In addition, the production of extremely high power optical pulses requires a chain of huge amplifiers to further intensify the emitted laser pulse. These areas of laser physics have generated many fascinating nonlinear mathematical models. The review article by Haken [9] develops in detail the problems that have arisen in the description of a laser, especially the problems of describing a collection of seemingly independent systems (the emitting atoms) which undergo an almost discontinuous transition to a state in which they behave collectively. A thorough treatment of the physical principles of lasers and amplifiers is contained in the book [10].

We first look at a mathematical description of an optical resonator. An optical resonator is essentially a pair of mirrors which provides a means for insuring the destructive interference of light waves which are out of phase with the rest of the electromagnetic energy. This principle of the destructive interference upon internal reflection is responsible for the iridescence patterns that oil on water displays (called Newton's rings). In a laser, a pair of mirrors can superpose all the coherent waves in the medium. There are two common types of resonators, which differ in the phase relations that they demand for constructive interference: the Fabry-Perot uses parallel mirrors which require that the total path length traversed by the light be constant; and the confocal resonator uses curved mirrors which allow for a focusing of the beam. Each resonator leads to an integral equation whose kernel is determined by the particular geometry designed to give constructive interference. Specifically, the prototype model is an integral equation of the Fredholm type:

$$\int_{-1}^{1} K(x,y)\ \psi(y)\ dy = \lambda\ \psi(x)$$

where

$\psi(x)$	is the amplitude of the radiation
y	is a coordinate perpendicular to the axis of the system
λ	is an eigenvalue
$K(x,y) = e^{ia(x-y)^2}$	for the Fabry Perot resonator
$K(x,y) = e^{2iaxy}$	for the confocal resonator.

This eigenvalue problem arises from the physical requirement that the amplitude of the laser radiation should replicate itself after each round-trip pass through the resonator. The mathematical problem of determining the eigenvalue spectrum is challenging because the integral operator is not self-adjoint and thus the eigenvalues are not real (and may not even exist!). However, a theorem of H. Hochstadt [11] guarantees the existence of an infinite set of eigenvalues for most problems of physical interest. Actual computation of the eigenvalues is of considerable importance to designers of laser systems because the imaginary parts of the eigenvalues give the loss of the open resonator. A detailed analysis of a simple example has recently been given by Cochran and Hinds [12]. There is need for further analysis of integral equations of this type in higher dimensions and in more complicated, and therefore more realistic, geometries. Cases where the imaginary parts of the eigenvalues are large are of special interest in coping with the problem of parasitic oscillations in laser amplifiers.

Once the laser beam has been produced and leaves the optical resonator, its propagation through various material media is described by a wave equation having the general form:

$$\frac{\partial^2 E}{\partial t^2} - \Delta E = P$$

where:

E	is the amplitude of the laser radiation
P	is the interaction of the radiation and matter
Q	is the state of the material medium

$$\frac{\partial P}{\partial t} = f(P,Q,E) \; , \qquad \frac{\partial Q}{\partial t} = g(P,Q,E) \; .$$

Now, in a vacuum, where there is no matter: $Q = P = 0$, and for a so-called linear dielectric:

$$P = \frac{\partial^2}{\partial t^2} E \; , \qquad \alpha = \text{a function of } Q.$$

These equations are a combination of Maxwell's equations describing the electric and magnetic fields in a vacuum and the "constitutive" equations for Q and P which describe the behavior of matter under the influence of these fields. The forms of the equations for P and Q are questions of physics and are currently very active fields of theoretical and experimental investigation. Needless to say, there is an abundance of models available, mostly nonlinear and/or stochastic, for which detailed mathematical analysis would be fruitful and beneficial to laser fusion research.

A simple but useful model for laser pulse propagation in an amplifier is provided by the "rate" approximation. In this approach, the phase relations are ignored, and only the modulus of the amplitude is studied. In this approximation (similar to the approximation of light waves by light rays) the following equation holds for $W = |E|^2$:

$$\frac{\partial W}{\partial z} + \frac{\partial W}{\partial t} = WQ \ , \qquad W(0,t) = W_0(t) \ ,$$

and

$$\frac{\partial Q}{\partial t} = - 2\ WQ \ , \qquad Q(z,-\infty) = Q_0(z) \ .$$

By changing the time variable, $\tau = t - z$, one obtains the equation

$$W_\tau + 2W^2 = f(\tau)W \ ,$$

where $f(\tau)$ is determined by the data. This nonlinear equation can readily be solved by the method of characteristics [13-14] yielding an explicit expression for W. Physically, the pulse of laser light is imagined, in this approximation, to propagate in the z-direction, changing in amplitude with the same dependence that it does in time. The prediction of this theory is that the laser pulse will propagate with a characteristic shape, rising very quickly in time and then falling off from its maximum relatively slowly. This general pulse shape, predicted in 1963, was later seen to create a problem for achieving laser fusion. It is known from computer simulations of the laser-fuel interaction that optimum energy transfer between the laser and the fuel pellet occurs in the case that the laser amplitude rises slowly and reaches its peak near the tail end of the pulse. Thus, simple propagation through an amplifier deforms the pulse in a way inimical to laser fusion, and the idea of "pulse shaping" becomes necessary. One method is to compose an overall burst of laser energy from a number (around ten) of much shorter pulses whose amplitudes are varied separately. Conditions under which the "rate" approximation is applicable is still an unsolved question, and further physical insights into the pulse-shaping problem depend heavily on the numerical solution of more complicated wave equations and on analytical studies of alternative methods [15-18].

Some of the real flavor of current nonlinear optics research is shown by an approach to the problem of beam propagation which begins to deal fully with the nonlinearity of the problem. This is the "quasi-optical" approximation and is important not so much for the exact results which it produces, but for the insight which it gives into the kind of nonlinearities implied by the high energy densities of laser fusion. In this approximation, the pulse of laser light is imagined to propagate with its approximate monochromaticity modified only slightly by a slowly-changing envelope. We define:

$$E = \text{Real}\ [\mathcal{E} e^{i(kz-\omega t)}]\ , \qquad \mathcal{E} = \mathcal{E}(x,y,z,t)\ ,$$

with similar expressions for P and Q. One then can straightforwardly derive the "quasi-optical" wave equation:

$$i(\mathcal{E}_z + \mathcal{E}_t) + \mathcal{E}_{xx} + \mathcal{E}_{yy} = \mathcal{P}(x,y,z,t)\ , \qquad \mathcal{P}_t = (\mathcal{P},\ \mathcal{Q},\ \mathcal{E})\ , \quad \text{etc.}$$

Notice that the assumption here is that the variation in the z-direction is "laser-like" and that interesting geometrical effects that are distinguished from its field-like propagation occur only in the directions traverse to the beam.

In the case of laser propagation in a material with "local" nonlinear index of refraction, the polarization has the form, $\mathcal{P} = \alpha(|\mathcal{E}|^2)\mathcal{E}$. Retaining only the leading term in expansion of the function α and restricting consideration to the one dimensional model, one obtains

$$i\mathcal{E}_t + \mathcal{E}_{xx} + 2|\mathcal{E}|^2\mathcal{E} = 0\ .$$

This equation is called the nonlinear Schrödinger equation (NLSE) and is one of a large class of nonlinear partial differential equations which can be attacked by the same methods [19]. In 1972 Zakharav and Shabat [20] showed that this equation can be linearized by using a technique called the "inverse scattering transform." Zakharov and Shabat discovered a pair of Hermitian operators:

$$L = \begin{pmatrix} i\partial x & -iE \\ iE^* & i\partial x \end{pmatrix}$$

$$B = \begin{pmatrix} -|E|^2 & -E_x \\ -E_x^* & |E|^2 \end{pmatrix}$$

which satisfy the commutation relation, first discussed by Lax [21],

$$iL_t = BL - LB \ .$$

The significance of these operators is seen by considering the eigenvalue problem for L:

$$L\phi = \lambda\phi \ .$$

It follows that

$$\frac{d\lambda}{dz} = 0 \qquad \text{and} \qquad i\phi_t = B\phi \ ,$$

i.e., we have an "isospectral flow". The method of solution is the following: we want to get from $E(x,0)$ to $E(x,t)$, but cannot solve the NLSE which gives this time evolution directly. Thus, with $E(x,0)$ we associate a $\phi(x,0)$. The evolution of $\phi(x,0)$ in time is then given by the linear equations given above. Having obtained $\phi(x,t)$ it is then possible to find the associated $E(x,t)$ by inverse scattering theory.

Thus, the inverse scattering method allows us to find the time evolution of the solution to the NLSE by solving a linear equation and two eigenvalue problems. Two very interesting consequences follow from this analysis of the NLSE:

(1) "Solitary wave" solutions: there is a class of solutions to the NLSE which are isolated lumps of electromagnetic energy, classically known as solitary waves. In addition, however, these waves manage to balance the dispersion in the wave equation against the nonlinearity in such a way that they propagate with an unchanging amplitude. This property is very striking but even more interesting is the fact that the NLSE admits multi-soliton solutions, in which the solitons can collide and reflect from each other without being distorted. They mimic the behavior of indestructible particles in their ability to maintain their identity [22]. In the physical situation of the propagation of laser light, these solitons correspond to

the formation of filaments of very intense light that are produced by the self-focusing of the beam. That is, the beam as it propagates (through either z or t) breaks up into a number of much smaller threads of light, each of which has an energy density thousands, even hundreds of thousands of times higher than the beam initially. Such filaments are observed experimentally, although no quantitative comparison of experiment and one-dimensional theory is yet possible.

(2) "Geometric" implications: in the past several years, the study of the inverse scattering method has given rise to a number of different approaches which have built on a geometric foundation. These theories have used differential geometry to understand the extraordinary fact that these nonlinear differential equations have these linear operators associated with them. One of the most interesting of these theories is the recent work of Lamb [23]. Using the fact that the motion of a vortex filament in a fluid is also described by a NLSE and that it can support soliton-like disturbances, Lamb was able to deduce the associated linear equations of Zakharov-Shabat. The key to this connection is the use of Riemannian geometry of the vortex (its curvature and torsion) to describe the soliton disturbance. This is the intuition that Riemann himself had concerning the connection between dynamics and geometry, that geometry is a dynamical quantity.

The mathematical challenges presented by laser fusion are probably more evident in the field of nonlinear optics and the study of nonlinear partial differential equations than in any other field. The single soliton solutions to several of these equations that arise in laser plasmas have been found in one dimension; however, neither the general multi-soliton solution in one dimension nor general solutions in more than one dimension have been found. This problem has withstood concerted attack for many years [24,25]. Even so, the attempts to solve this problem have opened up new areas in differential geometry, scattering theory, theory of integral transforms, and functional analysis. As in most of laser fusion physics many of the new ideas have been motivated by numerical solutions to the equations. In the case of self-focusing the initial observation of solitary solutions came from a one-dimensional computer solution and all the insight that is available into two-dimensional problems has come from similar solutions in two dimensions. At this point, however, solutions in x and t, and x, y and t, are available. The truly three-dimensional solutions are four dimensional on a computer and are practically unapproachable at present. The problems presented by these equations on a machine will be discussed in detail later, but even the preliminary computer work which has been done is largely responsible for the insight that now exists on these nonlinear partial differential equations.

PLASMAS

Once the laser energy reaches the fuel pellet and begins to interact with the matter in the pellet, the mathematics and physics present another set of problems. The general principle involved in the achievement of the conditions for thermonuclear fusion is the creation of an energy density sufficient to ignite fusion reactions in the fuel of the pellet. The conventional description of this condition for sufficient burn of the fuel to return the energy required for ignition is called the Lawson criterion:

$$n\tau > 10^{14} \text{ sec/cm}^3$$

where:

n = the particle density of the plasma

τ = the time of confinement at a temperature of $10^8 \circ C$.

This is a statement of the minimum product of time and energy density required for so-called "breakeven". When this requirement is recast in terms of the physics of the imploding pellet, it becomes a condition on the radius of the pellet (since $\tau \approx R/v$):

$$\rho R > .2 \text{ grams/cm}^2 , \qquad \rho = \text{mass density.}$$

The amount of laser energy required for ignition is roughly proportional to the mass of the pellet, so that:

$$E_L \propto M \propto \frac{1}{\rho^2} (\rho R)^3 \propto \frac{1}{\rho^2} .$$

This means that if the density of the pellet can be increased, the required laser energy decreases, and, in fact, decreases as the square of the density. A large amount of the experimental work in laser fusion at the present time centers around the problem of realizing such a compression, now envisaged to be on the order of a thousand times liquid density (about a thousand grams per cubic centimeter, or, a thousand pounds per pint). At these material and energy densities, the mathematical physics required has confronted physicists with a variety of challenging and still unsolved problems [26]. Here we look at a few of these: the idea of isentropic implosion, some of the nonlinear waves associated with this implosion, and numerical techniques for simulating the laser-plasma interaction and resulting hydrodynamic motions.

The simplest assumption to make in studying the compression of a small pellet of fuel is that the implosion and compression occur uniformly throughout the volume of the pellet. It is also a natural first approximation to assume that the compression occurs so fast that the heat generated travels more slowly than the compression; this kind of compression, called isentropic (since there is no heat transfer, hence, no entropy change), is certainly physically desirable since the work done by the compression is done against a relatively cold fuel rather than one that is hot. To get a rough idea of the sort of difficulties involved in an isentropic compression, Kidder [27-28] published the following simplified considerations. He began with the equations that approximate the behavior of the plasma as a fluid-like continuum:

$$\rho_t + \nabla\cdot(\rho\underline{v}) = 0$$

$$\underline{v}_t + \underline{v}\cdot\nabla\underline{v} = -\frac{1}{\rho}\nabla p$$

where

ρ is the density

$\underline{v}$ is the velocity

p is the pressure

and the experimental equation of state for isentropic changes in density:

$$p\rho^{\gamma} = \text{constant}$$

where:

γ is a constant experimentally determined, and approximately equal to five-thirds.

If we use Lagrangian coordinates (that is, a coordinate system which moves together with the fluid) and assume spherical symmetry, then we have a one-dimensional problem describing the time rate of change of the radius of the fuel pellet:

$$r(r_0,t) = r_0\, h(t)$$

$$v(r_0,t) = r_0\, \dot{h}(t)$$

$$\rho(r_0,t) = \rho_0(r_0)/h^3(t)$$

so that

$$h^{(3\gamma-2)}\, \ddot{h}(t) = -\frac{1}{t_0^2} = \text{const.}$$

This equation, although nonlinear, is easy to solve for the above value of γ and one obtains

$$h(t) = \left(1 - \frac{t^2}{t_0^2}\right)^{1/2},$$

and also a pressure history:

$$p(t) \propto \left(1 - \frac{t^2}{t_0^2}\right)^{-5/2}.$$

Most importantly, these equations allow a time history of the laser pulse to be prescribed. The following pulse shape is necessary to achieve isentropic compression:

$$\dot{E}(t) \propto R^2 p^{3/2} \propto \left(1 - \frac{t^2}{t_0^2}\right)^{-11/4}.$$

This equation describes a very rapidly rising pulse of laser energy, which in a real situation must rise four orders of magnitude in ten billionths of a second. This is a formidable technological problem, but, as described above, it has laid the basis for initial studies into the requirements for a workable laser system.

The actual design studies for laser energetics, however, have come out of sophisticated computer models of the interaction of the laser and plasma [29,30]. A few aspects of the numerical problems will be examined in the next section, but some of these more detailed interactions are exemplified by the nonlinear wave motions that the high energy density of the laser creates. The energy transport from the hot surface of the pellet where the laser energy is absorbed, into the cooler interior of the pellet can be studied using a nonlinear form of diffusion equation where the rate of diffusion depends on the energy density u:

$$\frac{\partial u}{\partial t} = \frac{\partial}{\partial x} u^n \frac{\partial}{\partial x} u \ , \qquad t \geq 0 \ .$$

As an example, we consider the spreading of energy from a point source of strength Q:

$$U(x,0) = Q\ \delta(x) \ .$$

We observe that this problem is invariant under the one parameter (L) continuous group of transformations

$$x' = x/L$$

$$t' = Q^n t/L^{n+2}$$

$$u' = Lu/Q \ .$$

This allows us to reduce the problem to that of solving the ordinary differential equation

$$(n + 2) \frac{d}{d\xi} (f^n \frac{df}{d\xi}) + \xi \frac{df}{d\xi} + f = 0 \ ,$$

where the similarity variable is given by

$$\xi = x/(tQ^n)^{1/(n+2)} \ .$$

It turns out that in this case we can solve this nonlinear equation exactly:

$$f(\xi) = \begin{cases} A\left[1 - \dfrac{\xi^2}{\xi_0^2}\right]^{1/2} , & \xi \leq \xi_0 \\ 0 , & \xi > \xi_0 \end{cases}$$

where $A = [n\xi_0^2/2(n+2)]^{1/n}$, and ξ_0 is a certain easily calculable numerical constant. We can then obtain the solution [31]:

$$U(x,t) = U_0(t)\left[1 - \frac{x^2}{x_f^2(t)}\right]^{1/n} , \qquad x_f(t) = \xi_0 (tQ^n)^{1/(n+2)} ,$$

where $u_0(t)$ is a known function. This shows that nonlinear thermal waves, unlike linear diffusive motions, can propagate with a sharply defined front located at $x = x_f(t)$. Other relevant applications of the similarity methods can be found in [31-33]. A large number of other techniques have been used to attack the differential equations that arise in describing various aspects of the laser-plasma interaction, including group theoretic methods, asymptotic methods, transform methods, and, rarely, classical special functions. However, in general, the analytic methods that are successful depend on taking the nonlinearity of the plasma and its high-energy density regimes as *fundamental*, and require an investigator to give up the linearized solutions to the differential equations even as a guide to the behavior of the nonlinearities. The examples of solitons and thermal waves described above are good case studies in this situation.

In a way that is true of laser plasma physics more than any other part of the physics of controlled fusion plasmas, numerical solutions to the dynamical equations have played a key role. Besides offering qualitative insight into equations that are otherwise insoluble, these computer programs allow the solution of exceedingly complex sets of equations which describe the most detailed physical picture that physicists have been able to provide. In other words, computational mathematics has progressed from the stage of solving idealized model equations to the stage where we are now able to solve the real equations describing the real physics, insofar as it is known.

The numerical simulation procedure begins with a set of physical processes which have been selected as the controlling factors in a given experimental situation. The physical laws determining the laser beam propagation and interaction, plasma dynamics, thermonuclear reactions, transport processes, properties of matter at high energy density, etc., must be formulated as a closed system of equations and in a form amenable to numerical solution. This last requirement often means that one must carry through a careful asymptotic analysis to discover the conditions under which a system of "microscopic" equations may be replaced by a simplified system of "macroscopic" equations. For example, the asymptotic transition from transport equations to diffusion equations has given rise to extensive mathematical investigations [34]. Having formulated the equations to be solved and having tabulated the appropriate data pertaining to material properties, one chooses the dimensionality of the numerical simulation taking into account the available symmetries, the geometries of the physical configurations, and the computational feasibility. Next, the equations describing continuous physical processes must be approximated by discrete analogs which allow for the finite capabilities of computers. For example, partial differential equations may be replaced by finite difference equations. To accomplish this in a manner which is faithful to the equations being solved and to the physical processes they represent is an arcane art which draws on the most advanced knowledge of the fields of numerical analysis and computational mathematics. The algorithm designer must take into account such factors as numerical stability, truncation and roundoff errors, computational efficiency (related to speed of convergence), versatility, and programming simplicity. A good discussion of many of the numerical techniques used in laser fusion simulation codes can be found in [35-36].

As an example of a powerful and widely used algorithmic concept, we consider the "splitting-up method" [36] as applied to the nonlinear diffusion equation

$$u_t = f(x,t,u) + \nabla\cdot(K(x,t,u)\,\nabla u) .$$

The basic idea of the splitting algorithm is to advance the solution in time using locally linearized operators to obtain a second order accurate approximation:

$$u(t + \Delta t) \approx e^{(\hat{F} + \hat{D})\Delta t}\, u(t)$$

$$\approx e^{\hat{F}\Delta t/2}\, e^{\hat{D}\Delta t}\, e^{\hat{F}\Delta/2}\, u(t) ,$$

where $\hat{F}$ and $\hat{D}$ are appropriate centered finite difference approximations to the operators

$$F = f_u(x,t,u) \quad , \text{ a multiplication operator}$$

$$D = \nabla\cdot(K(x,t,u)\,\nabla) \text{ , a 2nd order differential operator}$$

The error made in this approximation depends on the commutator of the operators $\hat{F}\,\Delta t$ and $\hat{D}\,\Delta t$, and will be small if Δt is small and if the solution is smooth. To proceed, one needs an approximation to the exponential operators, such as

$$e^{\hat{D}\Delta t} \approx \frac{1 + \hat{D}\,\Delta t/2}{1 - \hat{D}\,\Delta t/2} + O(\Delta t^3) ,$$

which gives rise to an implicit equation for $u(t + \Delta t)$, which in turn may be solved by iterative methods. This "splitting" method has the advantages of high accuracy, great flexibility, and efficient use of machine time and memory. There are many variations of this basic idea currently in use, and further analyses of these and other numerical methods are needed in laser fusion studies.

The final step in numerical simulation procedure is to integrate all of the physics, mathematics, and numerical algorithms into a concrete implementation of machine-readable instructions and data. The resulting computer code produces numerical and graphical output which can be compared directly to the results of experimental measurements. To write a computer code of such enormous complexity requires many years of sustained effort by many expert scientist-programmers. The programming requires great skill, ingenuity, and creativity, as well as much dogged perseverence.

It is commonly acknowledged that the most comprehensive and successful laser fusion simulation code now available is LASNEX, a code developed in recent years at Lawrence Livermore Laboratory [37,38]. Having been built up from earlier [39,40] hydrodynamic-transport codes, LASNEX is basically a code which solves the equations of compressible magnetohydrodynamics in two-dimensions (cylindrical symmetry) in Lagrangian coordinates. In addition, the code takes into account the mutual interactions of the laser light, the electrons of the plasma (which are themselves divided into two populations, thermal and suprathermal), the ionic species of the plasma, the photons which the plasma radiates (which are divided into several groups), the magnetic field created by currents flowing in the plasma, and the products of the nuclear fusion reactions -- mostly neutrons and alpha particles. All together, the code is designed to solve a system of equations involving vector unknowns and variable coefficients made up of 20 nonlinear integro-differential equations. Extensive tables of material properties (equations of state and opacities) provide the coefficients and other needed relations. A typical run with this code may consume about ten hours of machine time on a CDC 7600, one of the fastest and largest computers now available.

Despite the many successes of LASNEX, especially its ability to predict many of the results of present laser fusion experiments [8], there is need for still better codes which will be able to incorporate more refined and exact physical theories, and more accurate and efficient numerical methods [41,42]. The coming on-line of a new generation of computers, such as the CRAY I at LASL and the CDC STAR at LLL, provides a new opportunity for the construction of an ultra sophisticated laser fusion code which may finally be able to include all of the relevant physics in a fully three-dimensional numerical simulation.

The challenge of laser fusion has forced the scientists and physicists involved in its mastery into new areas of mathematics and physics, all of which revolve about the idea of nonlinearity. Some of the general approaches to the problem have been outlined, not in an exhaustive way, but rather to indicate the terrain on which the battle is being fought.

ACKNOWLEDGMENTS

Much of the information for this article was gathered while the author was a Visiting Staff Member at Los Alamos Scientific Laboratory. I am grateful for helpful discussions there with numerous staff members. Also I thank G. Zimmerman of the Lawrence Livermore Laboratory for information about the LASNEX computer program. Special thanks are due to S. Bardwell of the Fusion Energy Foundation for valuable assistance.

REFERENCES

1. J. Nuckolls, L. Wood, A. Thiessen, and G. Zimmerman, "Laser compression of matter to super-high densities: thermonuclear (CTR) applications," Nature 239 (1972), 139-142.
2. J. Nuckolls, J. Emmett, and L. Wood, "Laser-induced thermonuclear fusion," Physics Today, Aug. (1973), 46-53.
3. K. Boyer, "Laser initiated fusion -- key experiments looming," Astronaut. Aeronaut. 11 (1973), 28-38.
4. J. Emmett, J. Nuckolls, and L. Wood, "Fusion power by laser implosion," Sci. Amer. 230 June (1974), 24-37.
5. K. A. Brueckner and S. Jorna, "Laser-driven fusion," Rev. Mod. Phys. 46 (1974), 325-367.
6. "AEC (ERDA) Laser and Electron-Beam Programs FY 1976-1980," WASH 1363,

July 1974.

7. Los Alamos Scientific Laboratory Report "Laser fusion program" LA-6245-PR (1976).

8. Lawrence Livermore Laboratory Report "Laser program annual report -- 1975" UCRL-50021-75 (1976).

9. H. Haken, "Cooperative phenomena in systems far from thermal equilibrium and in nonphysical systems," Rev. Mod. Phys. 47 (1975), 67-121.

10. M. Sargent III, M. O. Scully, and W. E. Lamb, Jr., Laser Physics, Addison-Wesley, Reading, Mass. 1974.

11. H. Hochstadt, "On the eigenvalues of a class of integral equations arising in laser theory," SIAM Rev. 8 (1966), 62-65.

12. J. A. Cochran and E. W. Hinds, "Eigensystems associated with the complex-symmetric kernels of laser theory," SIAM J. Appl. Math. 26 (1974), 776-786.

13. R. Bellman, G. Birnbaum, and W. G. Wagner, "Transmission of monochromatic radiation in a two-level material," J. Appl. Phys. 34 (1963), 780-782.

14. L. M. Frantz and J. S. Nodvik, "Theory of pulse propagation in a laser amplifier," J. Appl. Phys. 34 (1963), 2346-2349.

15. G. L. Lamb, Jr., "Analytical descriptions of ultrashort optical pulse propagation in a resonant medium," Rev. Mod. Phys. 43 (1971), 99-124.

16. G. L. Lamb, Jr., "Amplification of coherent optical pulses," Phys. Rev. A12 (1975), 2052-2059.

17. R. A. Fisher and W. K. Bischel, "Numerical Studies of the interplay between self-phase modulation and dispersion for intense plane-wave laser pulses," J. Appl. Phys. 46 (1975), 4921-4934.

18. F. A. Hopf and P. Meystre, "Quantum theory of a swept-gain laser amplifier," Phys. Rev. A12 (1975), 2534-2548.

19. Lecture Notes in Mathematics, No. 515, ed. R. M. Miura, "Bäcklund transformations, the inverse scattering method, solitons, and their applications," Springer-Verlag, New York, 1976.

20. V. E. Zakharov and A. B. Shabat, "Exact theory of two-dimensional self-focusing and one-dimensional self-modulation of waves in nonlinear media," Zh. Eksp. Teor. Fiz. 61 (1971), 118-134 [Sov. Phys. JETP 34

(1972), 62-69].

21. P. D. Lax, "Integrals of nonlinear equations of evolution and solitary waves," Comm. Pure Appl. Math. 21 (1968), 467-490.
22. A. C. Scott, F. Y. F. Chu, and D. W. McLaughlin, "The soliton: a new concept in applied science," Proc. IEEE 61 (1973), 1443-1483.
23. G. L. Lamb, Jr., "Solitons and the motion of helical curves," Phys. Rev. Lett. 37 (1976), 235-237.
24. O. Svelto, "Self-focusing, self-trapping, and self-phase modulation of laser beams," Progress in Optics, Vol. 12 (ed. E. Wolf), North-Holland, 1974, pp. 1-51.
25. J. H. Marburger, "Self-focusing: theory," Prog. Quant. Electr. 4 (1975), 35-110.
26. Physics of High Energy Density, Course 48, ed. P. Caldirola and H. Knoepful, Academic Press, N. Y., 1971. See, especially, R. E. Kidder, "Interaction of intense photon and electron beams with plasmas," pp. 306-352.
27. R. E. Kidder, "The theory of homogeneous isentropic compression and its application to laser fusion," in Laser Interactions and Related Plasma Phenomena, Vol. 3, Plenum Press, 1974, pp. 449-464.
28. R. E. Kidder, "Laser compression of matter: optical power and energy requirements," Nucl. Fusion 14 (1974), 797-803.
29. J. S. Clarke, H. N. Fisher, and R. J. Mason, "Laser-driven implosion of spherical DT targets to thermonuclear burn conditions," Phys. Rev. Lett. 30 (1973), 89-92.
30. R. J. Mason and R. L. Morse, "Hydrodynamics and burn of optimally imploded deuterium-tritium spheres," Phys. Fluids 18 (1975), 814-828.
31. Ya. B. Zeldovich and Yu. P. Raizer, Physics of Shock Waves and High-Temperature Hydrodynamic Phenomena, Vol. II, Academic Press, N. Y., 1967, Chap. X, "Thermal Waves," pp. 652-684.
32. G. I. Barenblatt and Ya. B. Zeldovich, "Self-similar solutions as intermediate asymptotics," Ann. Rev. Fluid Mech. 4 (1972), 285-312.
33. D. D. Holm, "Symmetry breaking in fluid dynamics: Lie group reducible motions for real fluids," Los Alamos Scientific Laboratory Report LA-6392-T (1976).

34. G. C. Papanicolaou, "Asymptotic analysis of transport processes," Bull. Amer. Math. Soc. 81 (1975), 330-392.

35. R. D. Richtmyer and K. W. Morton, Difference Methods for Initial Value Problems, 2nd Ed., Interscience-Wiley, New York, 1967.

36. G. I. Marchuk, Methods of Numerical Mathematics, Springer-Verlag, N.Y., 1975.

37. G. B. Zimmerman, "Numerical simulation of the high density approach to laser fusion," Lawrence Livermore Laboratory Report UCRL-74811 (1973).

38. G. B. Zimmerman, and J. H. Nuckolls, "Theory of high density laser fusion," Lawrence Livermore Laboratory Report UCRL-76388 (1975).

39. J. A. Fleck, Jr., "The calculation of nonlinear radiation transport by a Monte Carlo method," Methods in Computational Physics, Vol. 1, Academic Press, N. Y. (1963), pp. 43-65.

40. W. D. Schulz, "Two-dimensional Lagrangian hydrodynamic difference equations," Methods in Computational Physics, Vol. 3, Academic Press, N. Y. (1964), pp. 1-45.

41. H. L. Wilson, "A new numerical technique for solving the time-dependent radiation transport equation," Progress in High Temperature Physics and Chemistry, Vol. 5 (ed. C. A. Rouse), Pergamon Press, N. Y. (1973), pp. 125-156.

42. F. D. Tappert, "Diffractive ray tracing of laser beams," J. Opt. Soc. Amer. 66 (1976), 1368-1373.

Courant Institute of Mathematical Sciences
New York University
New York, New York 10012

MATHEMATICAL PROBLEMS IN MODELLING ENERGY PRODUCTION AND DISTRIBUTION

Proceedings of Symposia in Applied Mathematics
Volume 21
1977

ESTIMATION OF UNDISCOVERED OIL AND GAS

E. Barouch and G. M. Kaufman[1]

ABSTRACT. What additions to recoverable oil and gas will accrue from an increment of exploratory effort? To answer this question we first construct a probabilistic model of oil and gas discovery from assumptions that describe the manner in which exploration technology and observed statistical regularities of the size of pools interact to generate discoveries. The basic assumptions incorporate certain geological facts and the content of a variety of statistical studies of the discovery process.

The parameters of the model are in general not known with certainty a priori. We study the joint density of discovery sizes and the (conditional) moments of future discovery sizes given what has been observed as a prelude to statistical estimation of these parameters. For the case of greatest practical interest - lognormal pool size distributions - we compute uniform asymptotic expansions for (1) the joint density of discovery sizes and for (2) the conditional moments of future discovery sizes. Approximate maximum likelihood estimation can be done using (1) and non-linear least squares estimation can be done using (2). Once the parameters are estimated, a predictive forecast of future discovery sizes can be generated.

1. INTRODUCTION. Petroleum exploration is random in character. A typical exploration program in a petroleum basin begins with a sequence of information gathering activities such as geological basinal analysis, geophysical reconnaissance surveying, detailed geophysical surveying, and well data analysis when available. The purpose of these activities is to identify prospects. A prospect is a geological configuration perceived by the geologist to have trapped hydrocarbons that forms a target for drilling. Given the current state of technology, the only way to know for certain whether or not a prospect contains hydrocarbons is to drill a well into it. In 1974 only one out of four exploratory wells drilled in the U. S. onshore and offshore (excluding Alaska) was successful. Only one out of eight wildcat wells - wells drilled with the intent of discovering new hydrocarbon pools or fields was an economic success. Exploration is risky indeed! The

---

AMS(MOS)subject classification (1970). Primary 60E05

[1]Research supported by the National Science Foundation under grant SIA74-22773.

sizes of discoveries made by this drilling effort, when measured in barrels of recoverable oil, vary by many orders of magnitude depending on geographical location and a host of geological and economic variables.

Drilling successes and failure data and size of discovery data possess certain statistical regularities which can be used to build a stochastic model of the discovery process whose aim is to provide probabilistic answers to two questions: first, within a given geographical area, how many undiscovered petroleum pools remain to be discovered and what is their size distribution? Second, what additions to known reserves will accrue from an increment of exploratory effort?

There is a practical need for credible answers computed from a model whose assumptions can be validated by statistical tests: the answers are an essential ingredient of a "correct" economic supply function for petroleum. Many current policy issues hinge on forecasts of the magnitude of our undiscovered petroleum resources and the rate at which they will be discovered and produced. For example, the timing for and economic viability of alternatives to petroleum are critically dependent on the amounts of undiscovered oil and natural gas in the U. S. and on costs of discovery and production.

To this end we have constructed such a model for individual petroleum zones or plays. A play is a collection of geographically close prospects with similar geological configurations; i.e. prospects located in the same geological basin all of which are of similar geological type.

In what follows we address these points.

(1) How data describing the size, measured in barrels of oil, of oil field discoveries lead to two basic assumptions about the discovery process from which a model of it can be built.

(2) What these assumptions imply.

(3) What computations must be done to estimate model parameters from observed data.

(4) How forecasts of future discovery sizes can be generated from the model.

2. FIELD SIZE DATA. Empirical studies show that relative frequency histograms of the sizes of discovered mineral deposits have large positive skewness; i.e. there tends to be a large number of deposits of "small" size and a small number of "large" size. This is true of oil and gas field sizes in particular. For many well explored petroleum plays, the relative frequency histogram of field or pool sizes closely approximates a lognormal density; i.e. the probability density function of a random variable, the logarithm of which possesses a normal or Gaussian probability law. (cf. [1], [6], [8]). For example, Figures 1a and 1b are graphs of the sizes of oil pools plotted on lognormal probability paper.* Figure 1a is composed from Leduc reef pools in Alberta province. In their study [1] of sizes of cretaceous fields in the Denver-Julesburg basin, Arps and Roberts classify 338 fields into fifteen size groups. Figure 1b displays the midpoints of these size groups on lognormal probability paper.

Studies such as these lend empirical support to the assumption that the size distribution of petroleum deposits <u>in nature</u> is lognormal. However, plays possessing a substantial number of prospects that have not yet been drilled have empirical cumulative functions of <u>discovery sizes</u> with "fatter than lognormal" right tails, i.e. more positively skewed distributions. Graphs of the empirical cumulative functions of two such plays - Keg River and Lower Manville, both in the Canadian western sedimentary basin - are displayed in Figures 2a and 2b. The "bend" upwards in the right tail is clearly discernable in both graphs.

Another feature of the data reflects petroleum folklore and follows naturally from the way in which exploration is conducted: "The big ones are usually discovered first - not always, but usually." Given a set of prospects whose elements have a common structural configuration and are located in the same geographical region, the bigger ones are most likely to be identified first. Economics usually dictates that a "larger" prospect will be drilled before a "smaller" prospect. Figure 3 is the graph of the empirical mean of (rescaled) sizes of pools in Alberta plays presented in order of discovery. Values of this function were computed by first dividing pool sizes within each play by an estimate of the mean size of pools within that play, then averaging the rescaled sizes of the first discovery in each play across all plays,

* Lognormal probability paper is designed so that the graph of the cumulative function of a lognormal probability law appears as a straight line on it. The closer to a straight line the empirical cumulative function of a data sample plotted on this paper, the more accurately it is approximated by a lognormal distribution.

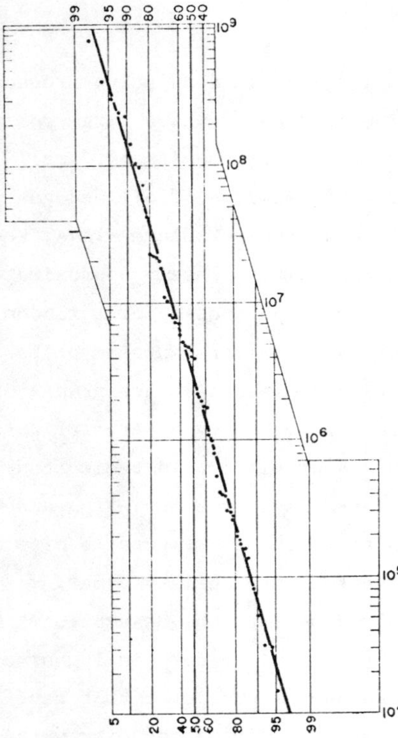

Fig. 1a

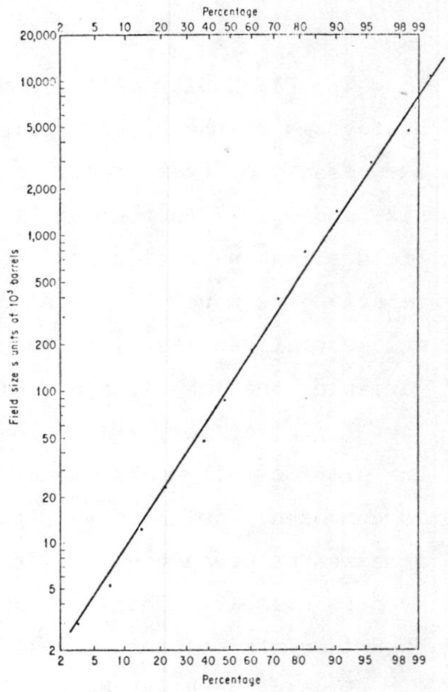

Estimated primary ultimate oil recovery from cretaceous fields in the Denver-Julesburg basin as of 1958

Fig. 1b

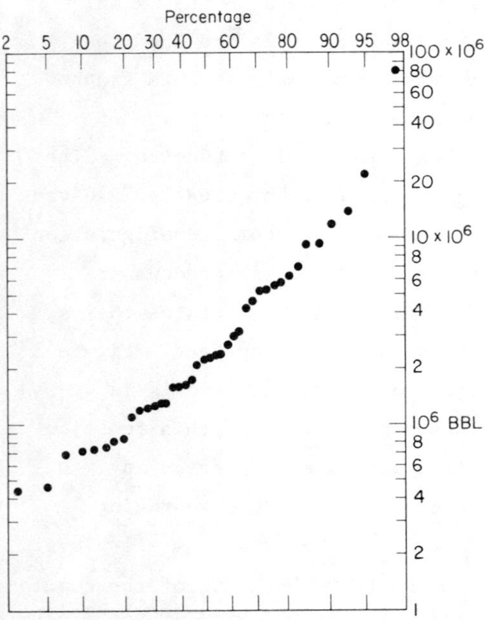

Fig. 2b

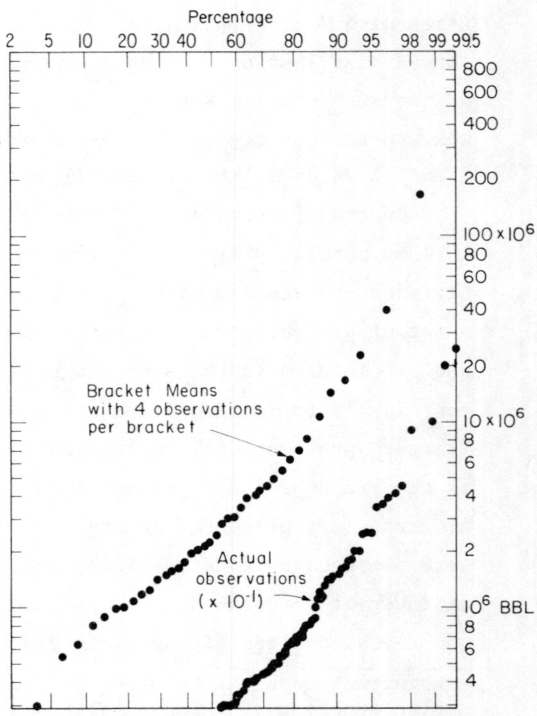

Fig. 2a

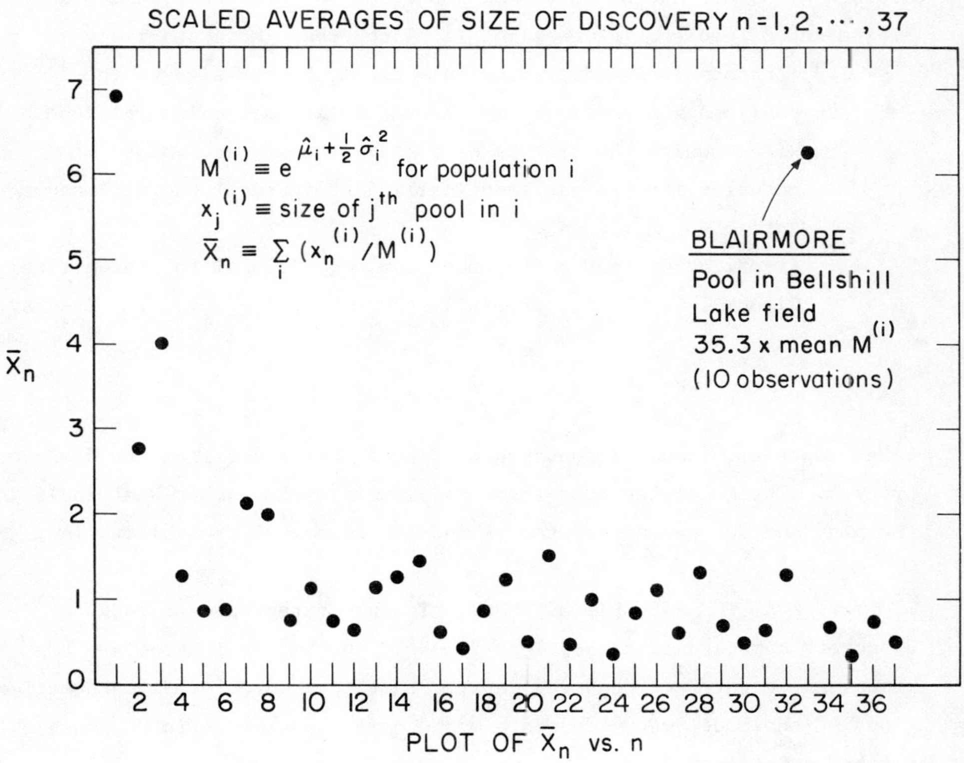
SCALED AVERAGES OF SIZE OF DISCOVERY n = 1, 2, ⋯, 37
$M^{(i)} \equiv e^{\hat{\mu}_i + \frac{1}{2}\hat{\sigma}_i^2}$ for population i
$x_j^{(i)} \equiv$ size of jth pool in i
$\bar{X}_n \equiv \sum_i (x_n^{(i)}/M^{(i)})$
BLAIRMORE
Pool in Bellshill
Lake field
35.3 x mean $M^{(i)}$
(10 observations)
$\bar{X}_n$
0
1
2
3
4
5
6
7
2 4 6 8 10 12 14 16 18 20 22 24 26 28 30 32 34 36
PLOT OF $\bar{X}_n$ vs. n

Fig. 3

second discoveries, etc.

3. BASIC ASSUMPTIONS. These features of the data suggest two assumptions, one that describes the size distribution of pools as deposed by Nature and another than describes the process by which a finite set of pools of given sizes is discovered. More precisely:

I. (Lognormal size distribution) Let A_i be the size of the ith pool among N pools deposed by nature in the geological zone within which the play takes place. The A_is are values of mutually independent identically distributed lognormal random variables, $\tilde{A}_1,\ldots,\tilde{A}_N$.

II. (Sampling without replacement and proportional to random size) Given $A_1,\ldots,A_N$ the probability of observing $A_1,\ldots,A_n$ in that order is

$$\prod_{j=1}^{n} A_j/(A_j+\ldots+A_N).$$

Taken together, I and II constitute a parsimonious model of the discovery process. The model has only three parameters: the number N of pools in the play and two parameters that index the lognormal probability law.

4. WHAT THE ASSUMPTIONS IMPLY. To what extent does a model of discovery composed of assumptions I and II in combination capture the principal features of observed data? Prior to a more formal mathematical study of the model we performed a monte carlo simulation in response to this question.*

Given a lognormal "superpopulation" with parameters $E(\log_e \tilde{A}_i) \equiv \mu = 6.0$ and $Var(\log_e \tilde{A}_i) = \sigma^2 = 3.0$ the discovery process was replicated 1000 times for each of five different values of N, the number of pools in a hypothetical play. If discovery sizes are generated according to assumptions I and II, it is obvious that they are neither independent nor identically distributed random variables. Suppose that we incorrectly assume that they are independent and identically distributed sample observations from a lognormal process. When plotted on lognormal probability paper, the expectation of the empirical cumulative distributive function** (empirical cdf) of a sample of n discovery sizes is close to

* See [7] for a more detailed description.

** As approximated by 1000 monte carlo trials.

linear in the interquartile range but exhibits a fatter right tail than that of a lognormal distribution function, a principal feature of the empirical cdfs for the Keg River and Lower Manville plays as displayed in figures 1a and 1b. The graph of the expectation of the simulated cdf for $n < N$ discoveries lies above the corresponding graph of the cdf for the lognormal distribution of Assumption I and is "tilted" as shown in Figure 4a. As $n \to N$, the former cdf approaches the latter.

Figure 4b displays graphs of a monte carlo approximation to the marginal expectation of the size of the nth discovery, $n = 1,2,\ldots$ for fixed $N = 100$, 150, 300, 600 and 1200. For fixed N, the marginal expectations of the first few discoveries are an order of magnitude loarger than the mean of the underlying lognormal distribution of assumption I. Regarded as a function of n, the expectations decline rapidly at first, pass through a turning point (at roughly $n \sim \sqrt{N}$), and then decline slowly; i.e. the mathematical form of the expectations changes character in moving from small n through the turning point to larger n.

These graphs mimic the empirical data as displayed in Figure 3.

While more sophisticated statistical tests of the validity of the basic assumptions are possible (see [2] for a brief précis), essential features of actual discovery size data are in fact captured by the model as posed here.

4. ANALYSIS OF THE MODEL. In order to do statistical testing of model structure, to estimate parameters of the mdoel from observed data, and to generate probabilistic forecasts of the sizes of discoveries yet to be made, we need a computable representation for the probability density of observed discovery sizes. To this end we have done an asymptotic study of this density and its moments.

Let f be a member of a class of densities concentrated on $[0,\infty)$ and indexed by a parameter $\theta \in \Theta$ and assume that each $\tilde{A}_i$ as defined in assumption I has density $f(\cdot|\theta)$. Define Y_j as the size of the jth discovery (jth observation) and $\underline{Y} = (Y_1,\ldots,Y_n)$ as a vector of discovery sizes in a sample of size $n \leq N$. Given θ, N, and infinitesimal intervals $dY_1,\ldots,dY_n$, assumptions I and II imply that the probability of observing $\tilde{Y}_1 \in dY_1,\ldots,\tilde{Y}_n \in dY_n$ in that order is, letting $b_j = Y_j+\ldots+Y_n$,

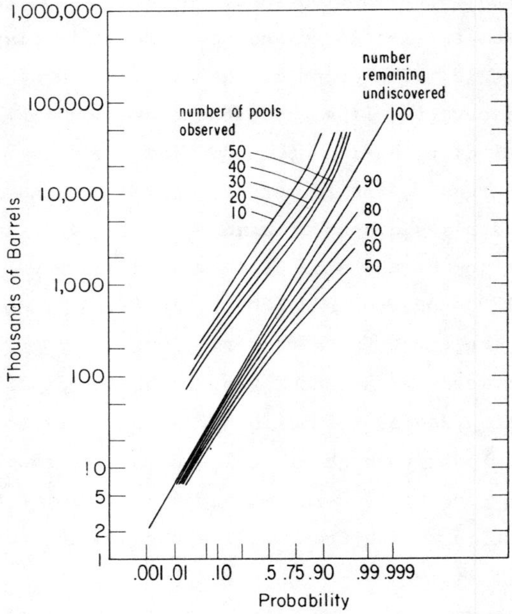

Simulated cumulative distribution functions for observed pool sizes and for pools remaining undiscovered when $N = 100$.

Fig. 4a

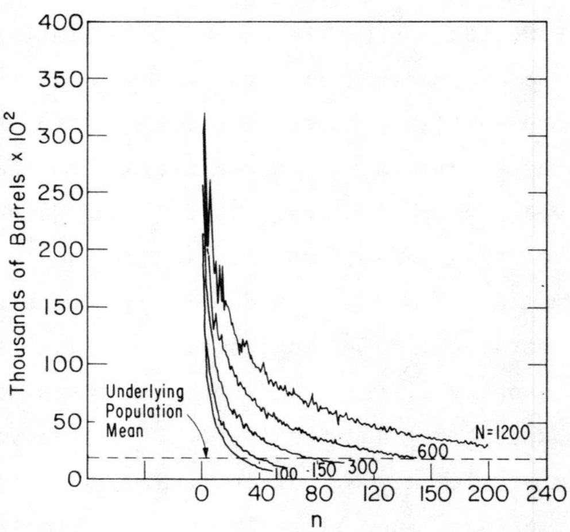

Simulated means of size of nth pool discovered for finite population sizes $N = 100, 150, 300, 600$, and 1,200.

Fig. 4b

$$P\{\underline{\tilde{Y}} \in d\underline{Y} \mid \theta, N\} = \frac{\Gamma(N+1)}{\Gamma(N-n+1)} \prod_{j=1}^{n} Y_j f(Y_j \mid \theta) dY_j$$
$$\times \int_0^\infty \cdots \int_0^\infty \prod_{j=1}^{n} [b_j + A_{n+1} + \ldots + A_N]^{-1} \prod_{k=n+1}^{N} f(A_k \mid \theta) dA_k \tag{1}$$

In the special case when f is an exponential density, $\lambda\exp\{-\lambda Y\}$, for $\lambda > 0$ and $Y \in [0,\infty)$, and zero otherwise, (1) can be evaluated exactly. In [2] it is shown that the density of $\underline{\tilde{Y}}$ can be expressed in terms of a linear combination of incomplete gamma functions and that the mean and variance of the size $\tilde{Y}_n$ of the nth discovery are $E(\tilde{Y}_n) = \frac{2}{\lambda}[1 - \frac{n}{N+1}]$ and $\mathrm{Var}(\tilde{Y}_n) = \frac{2}{\lambda^2}[1 - \frac{n(2N+1)}{(N+1)(N+2)} + \frac{n^2(N-1)}{(N+1)^2(N+2)}]$ respectively.

When the density of the $\tilde{A}_i$s is f and f possesses moments of all order, an asymptotic expansion of the density of $\underline{\tilde{Y}}$, valid for large $p \equiv N-n$ and fixed $b_j = Y_j + \ldots + Y_n$, $j = 1,2,\ldots,n$ is

$$I_{N,n}(\underline{Y}) \prod_{j=1}^{n} Y_j f(Y_j)$$

where, defining M_i as the ith moment of f and V as its variance,

$$I_{N,n}(\underline{Y}) = \frac{\Gamma(p+n+1)}{\Gamma(p+1)} \prod_{j=1}^{n} [pM_1 + b_j]^{-1}$$
$$\times \{1 - \tfrac{1}{2}pVg_2(pM_1,\underline{Y}) + \tfrac{1}{8}p^2V^2g_4(pM_1,\underline{Y}) \tag{2}$$
$$+ p(\tfrac{1}{6}M_3 - \tfrac{1}{2}M_1M_2 + \tfrac{1}{3}M_1^3)g_3(pM_1,\underline{Y}) + O(p^{-3})\}$$

The functions $g_m(pM_1,\underline{Y})$ are rational functions of $\underline{Y}$ and pM_1 and $g_m = O((pM_1)^{-m})$. The expansion (2) is valid in a region for which nV/pM_1^2 is smaller than order one.

When f is lognormal with parameters μ and σ^2, $M_1 = \exp\{\mu + \frac{1}{2}\sigma^2\}$ and $V = M_1^2[\exp\{\sigma^2\} - 1]$, so that $n\exp\{\sigma^2\}/p$ must be smaller than order one for (2) to be valid. A crude statistical analysis of plays in western Canada shows σ^2 to vary from .8 to 8.0 between plays, while n varies from 2 to 3 to nearly 200 between plays. Consequently, an expansion of the form (2) is not sufficient to cover all cases of practical interest; i.e. $n\exp\{\sigma^2\}/p$ order one, and $n\exp\{\sigma^2\}/p$ larger than order one.

We have computed asymptotic expansions of (1) for large p in these cases:

(i) n and $\exp\{\sigma^2\}$ fixed (formula (2)),

(ii) $\exp\{\sigma^2\}$ fixed and $n = O(\sqrt{p})$ or equivalently n fixed and $\exp\{\sigma^2\} = O(\sqrt{p})$,

(iii) $n\exp\{\sigma^2\}/p = O(1)$.

Cases (ii) and (iii) have been done by the method of steepest descent with two distinct steepest descent points that coalesce when $n\exp\{\sigma^2\}/p = O(1)$.

More particularly, for fixed μ,

$$\begin{aligned} M_1 &= \exp\{\mu + \tfrac{1}{2}\sigma^2\} = O(p^{1/4}) \\ K &= \frac{1}{n}\sum_{j=1}^{n} b_j = O(p^{5/4}) \\ V &= M_1^2(\exp\{\sigma^2\}-1) = O(p) \\ n &= O(\sqrt{p}) \\ e^{\sigma^2} &= O(\sqrt{p}) \end{aligned} \tag{3}$$

upon approximating the characteristic function

$$G(y) = \frac{1}{\sigma\sqrt{2\pi}} \int_0^\infty \exp\{-iyx - \frac{1}{2\sigma^2}\log^2 x\}\frac{dx}{x} \tag{4}$$

of the lognormal density by the first three terms in a moment expansion of it*, the density of $\tilde{\underline{Y}}$ is approximated by

$$\begin{aligned} & i^n n Q^n \binom{p+n}{n} \prod_{j=1}^{n} Y_j f(Y_j|\mu,\sigma^2)/[b_j + pM_1] \\ & \times \int_0^\infty \exp\{Q[\delta\log z - iz - \tfrac{1}{2}z^2]\}dz \end{aligned} \tag{5}$$

with

$$Q = \frac{(K+pM_1)^2}{pV} = O(\sqrt{p}) \quad \text{and} \quad \delta = n-1/Q = O(1)$$

Case (iii) corresponds to Q large and fixed δ of order one. As approaches 1/4, we have two coinciding saddle points and consequently a turning point at $\delta = 1/4$. When the two saddle points are close ($\delta \cong 1/4$) the contribution from both must be taken into account. The

* The parameter μ plays no essential role so we set it equal to zero.

usual method of steepest descent is not then applicable and we employ a more sophisticated scheme for asymptotic computation designed by Chester, Friedman, and Ursell [4].

Their method for constructing uniform expansions for the case of two nearly coincident saddle points begins by introduction of a new complex variable implicitly defined in terms of z as

$$h(z,\delta) \equiv \delta \log z - iz - \frac{1}{2}z^2 = \frac{1}{3}u^3 - \xi(\delta)u + A(\delta),$$

the parameters $\xi(\delta)$ and $A(\delta)$ being determined by the condition that the (u,z) transformation is uniformly regular, here near $\delta = 1/4$ and $z = -i/2$. They show that with $\xi(\delta)$ and $A(\delta)$ so computed, there is one branch of the transformation that is uniformly regular and use this fact to compute uniformly asymptotic expansions of $\int \exp\{-ph(z,\delta)\}dz$ of the form

$$\exp\{-pA(\delta)\}\{\frac{Ai(p^{2/3}\xi)}{p^{1/3}} \sum_{j=0}^{\infty} \frac{a_j(\delta)}{p^j} + \frac{Ai'(p^{2/3}\xi)}{p^{2/3}} \sum_{j=0}^{\infty} \frac{c_j(\delta)}{p^j}\} \tag{6}$$

where

$$Ai(\xi) = \frac{1}{2\pi i} \int_{\infty e^{-\pi i/3}}^{\infty e^{\pi i/3}} \exp\{\frac{1}{3}u^3 - \xi u\}du$$

and $Ai'(\xi) = dAi(\xi)/d\xi$; i.e. $Ai(\xi)$ is the Airy function and $Ai'(\xi)$ its derivative.

When n is larger than $O(\sqrt{p})$ the major contribution to the integral (5) comes from a region for which an approximation of the characteristic function G(y) of the lognormal distribution by $1 - iM_1 y - \frac{1}{2}M_2 y^2$ breaks down. A more delicate approximation scheme is needed. In order to design it, G(y) must be understood throughout its entire domain, including its asymptotic region. Approximations to G(y) for small argument, argument order one, and large argument are given in [3]. For very small y, G(y) can be approximated by a polynomial composed of the first few terms of its moment expansion; however, the polynomial disappears as y increases. In its asymptotic region,

$$G(y) \simeq \frac{1}{\sigma\sqrt{2\pi}} \exp\{-\frac{1}{2\sigma^2}\log^2 y\}\Gamma(\frac{\log y}{\sigma^2})\exp\{-\frac{1}{2\sigma^2}\psi^2(\frac{\log y}{\sigma^2})\}, \tag{7}$$

with $\psi(x) = \Gamma'(x)/\Gamma(x)$.

5. BEHAVIOR OF THE LIKELIHOOD FUNCTION. The likelihood function for μ, σ^2, and N is the density of $\underline{Y}$ at $\tilde{\underline{Y}} = \underline{Y}$ regarded as a function with arguments μ, σ^2, N, which we write as $L(\mu,\sigma^2,N|\underline{Y})$. Using the TROLL system at the National Bureau of Economic Research's Cambridge Computation Center, we designed a graphics routine that draws iso-contours of approximations to log L in μ and σ^2 for fixed N and $\underline{Y}$.

Figure 5 is an example. The vector $\underline{Y}$ has as elements sizes measured in 10^5 barrels of recoverable oil of the first 24 pools discovered in the Leduc play located in the Canadian western sedimentary basin. The number of pools in the play is fixed at N = 60. The iso-contours are smooth and appear almost elliptical in shape. There is a unique maximum at $\mu = 4.14$ and $\sigma^2 = 1.51$ and $24 \exp\{1.51\}/36 = 3.02 = O(1)$, so these parameter values lie in the uniform region. The function L is rather flat in the neighborhood of $\mu = 4.14$ and $\sigma^2 = 1.51$.

ISO-CONTOURS OF UNIFORM EXPANSION FOR $\log_e L(\mu,\sigma^2,N|\underline{Y})$ at N = 60, n = 24

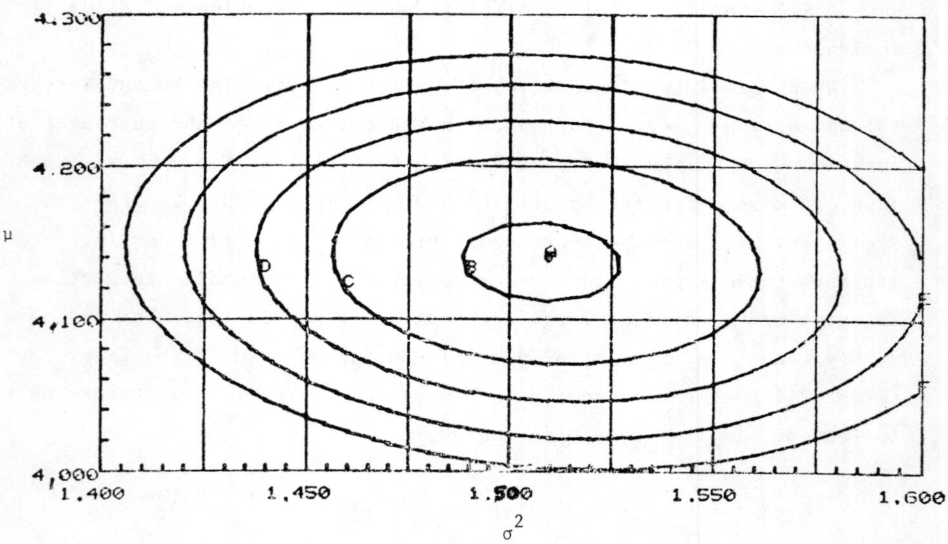

A=-49.0136, B=-49.02, C=-49.06, D=-49.10, E=-49.15, F=-49.20

Fig. 5

For fixed $\underline{Y}$ and each value of N displayed in Table 1 below there is a unique maximum for the uniform approximation to L at $\mu = \hat{\mu}(N)$ and $\sigma^2 = \hat{\sigma}^2(N)$. For n=24 and N varying from 43 - the number of pools discovered to date - to 1000, $n\exp\{\hat{\sigma}^2(N)\}/N-n$ varies from 5.28 to .53.

Table 1

N	$\hat{\mu}(N)$	$\hat{\sigma}^2(N)$	$\log_e L$	$n\exp\{\hat{\sigma}^2(N)\}/N-n$
43	4.21	1.43	-48.2	5.28
50	4.17	1.46	-48.6	3.97
60	4.14	1.51	-49.0	3.02
75	4.10	1.56	-49.7	2.24
100	4.05	1.67	-50.7	1.68
150	4.01	1.84	-52.9	1.20
200	3.97	1.98	-54.8	.99
400	3.91	2.39	-60.7	.70
600	3.89	2.67	-64.9	.60
1000	3.86	3.07	-70.6	.53

If, however, μ and σ^2 are held fixed, $L(\mu,\sigma^2,n|\underline{Y})$ regarded as a function of N with domain $[n,\infty)$ is sharply peaked and has a single maximum. We have computed this function numerically for only one case thus far: a sample of sizes of 59 North Sea discoveries and values 5.78 and 1.38 of μ and σ^2 respectively that maximize L for N fixed at 300; i.e. $\hat{\mu}(300) = 5.78$ and $\hat{\sigma}^2(300) = 1.38$. The function $L(5.78, 1.38, N|\underline{Y})$ of N peaks at N = 99. Details of this computation are presented in [5] where it is shown that (1) can be expressed in the form

$$\int_0^\infty Z(\lambda)[H(\lambda)]^{N-n}d\lambda \tag{8}$$

where $H(\lambda)$ is the Laplace transform of the lognormal density and

$$\begin{aligned} Z(\lambda) &= \frac{1}{2\pi i}\int_{-i\infty}^{i\infty} \exp\{\lambda z\} \prod_{j=1}^{n} [b_j + z]^{-1}dz \\ &= \sum_{j=1}^{n} p_j \exp\{-\lambda b_j\} \end{aligned} \tag{9}$$

with $p_j = \prod_{\substack{k=1\\k\neq j}}^{n} [b_k - b_j]^{-1}$. In order to account for the large number of cancellations of $Z(\lambda)$ near $\lambda = 0$, we computed it to 100 significant digits on the MIT Math Lab system.

Approximations to the first moments of sizes $\tilde{Y}_{n+1},\ldots,\tilde{Y}_N$ of as yet unobserved discoveries given sizes $Y_1,\ldots,Y_n$ of the first n discoveries have been computed. A numerical example is given below. We are currently studying the correlation structure of the $\tilde{Y}_j$s. Once it is understood, non-linear least squares methods may be used to compute approximate point estimates of μ, σ^2, and N.

6. FORECASTING - AN EXAMPLE. Oil supply from countries outside the OPEC cartel core is a current topic of research by members of the M.I.T. World Oil project. One important source of future supply for Western Europe is the North Sea. Using data on 59 North Sea field sizes and a subjective point estimate of N = 300 provided by an exploration geologist, we computed maximum likelihood estimates $\hat{\mu}(300) = 5.77$ and $\sigma^2(300) = 1.38$. This corresponds to a mean of 645 (in 10^6 barrels of recoverable oil) for f. Approximations to expectations of sizes of the sixtieth through the ninety-sixth discoveries given sizes of the first 59 discoveries are displayed in Figure 6.

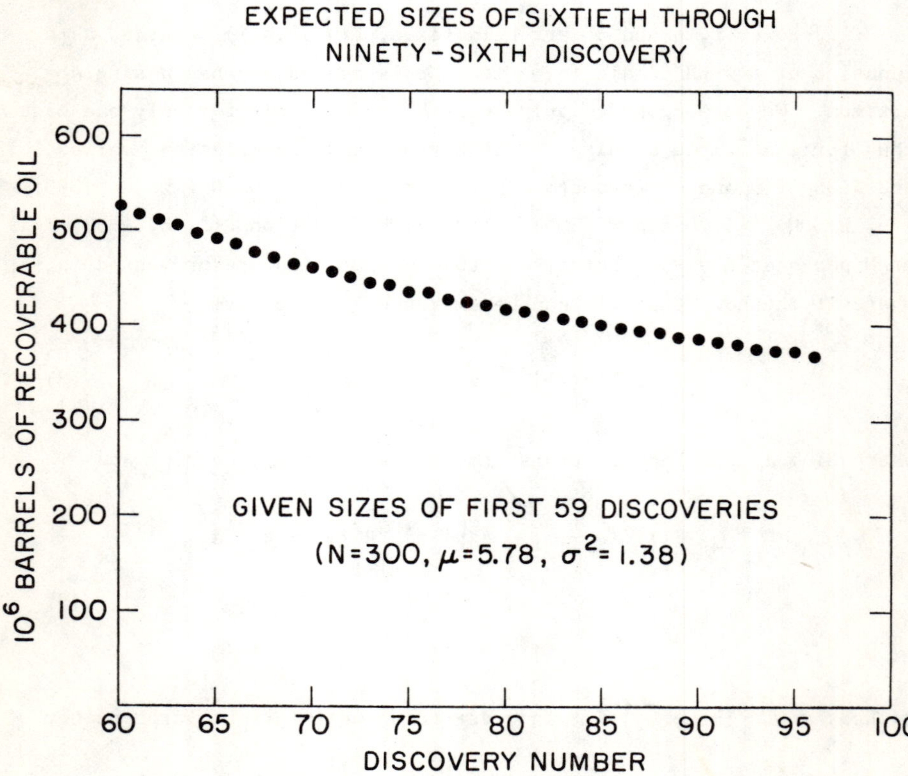

Fig. 6

REFERENCES

1. Arps, J. J. and T. G. Roberts (1958) Economics of Drilling for Cretaceous Oil Production on the East Flank of the Denver-Julesberg Basin, Bull, AAPG, Vol. 42, No. 11, pp. 2549-2566.

2. Barouch, E. and G. Kaufman (1976) "Oil and Gas Discovery Modelled as Sampling Proportional to Random Size:, M.I.T. Sloan School of Management Working Paper.

3. ________________ (1976) "On Sums of Lognormal Random Variables", M.I.T. Sloan School of Management Working Paper.

4. Chester, E., B. Friedman, and F. Ursell (1956) "An extension of the method of steepest descents", Proc. Camb. Phil. Soc. Vol. 53, pp. 599-611.

5. Barouch, E., Fateman, R., and Kaufman, G., "Computational Techniques in Modelling Oil and Gas Discovery" (In preparation)

6. Kaufman, G. (1962) Statistical Decision Theory and Related Techniques in Oil and Gas Exploration, Prentice-Hall, Englewood Cliffs, N.J.

7. __________, Balcer, Y., Kruyt, D. (1975) Studies in Geology No. 1, AAPG, pp. 113-142.

8. McCrossan, R.G. (1969) "An Analysis of Size Frequency Distribution of Oil and Gas Reserves of Western Canada", Canadian Journal of Earth Sciences, Vol. 6, No. 2, pp. 201-211.

ACKNOWLEDGEMENTS. We thank Joel Moses and Richard Fateman for the use of the Laboratory for Computer Science MACSYMA System which is supported by ERDA under contract #e(11-1)-3070.

CLARKSON COLLEGE OF TECHNOLOGY
and
MASSACHUSETTS INSTITUTE OF TECHNOLOGY

MASSACHUSETTS INSTITUTE OF TECHNOLOGY

Proceedings of Symposia in Applied Mathematics
Volume 21
1977

ON A PILOT LINEAR PROGRAMMING MODEL FOR ASSESSING PHYSICAL IMPACT ON THE ECONOMY OF A CHANGING ENERGY PICTURE**

George B. Dantzig* and S. C. Parikh*

ABSTRACT. This papers deals with a dynamic linear programming model on a pilot scale that attempts to describe in physical terms many of the technological interactions within and across the sections of the American Economy, including a detailed energy sector. The general aim of the model is to provide information on what the country could achieve over the long run in physical terms in the face of a changing energy picture.

The model is expected to (i) incorporate most recent available data of sufficiently good quality (albeit in an aggregated form) that it can be used to generate some meaningful scenarios showing how the economy might be affected if the energy picture evolves in a specified way, and (ii) provide us at the Systems Optimization Laboratory a prototype for research in solving large-scale linear programming models of energy systems.

1. INTRODUCTION. This paper reports on some of the ongoing work on models of energy systems at the Institute for Energy Studies and the Systems Optimization Laboratory of the Stanford University. It deals with a dynamic, linear programming model on a pilot scale that attempts to describe in physical terms many of the technological interactions within and across the sectors of the American Economy, including a detailed energy sector. The general aim of the model is to provide information on what the country could achieve over the long term (say 30 years) in physical terms in the face of changing energy picture.

---

*Research partially supported by the Office of Naval Research under Contracts N00014-75-C-0865 and N00014-75-C-0267; the Energy Research and Development Administration Contract AT (04-3)-326 PA #18; and National Science Foundation Grant MPS71-03341 A03.

**Reprinted with minor revisions from ENERGY: Mathematics and Models with permission of SIAM. Copyright 1976 by Society for Industrial and Applied Mathematics, Philadelphia.

Mathematical programming models that link activities of the economic sectors with those of a detailed energy sector and describe interactions over time can provide comprehensive and effective means for evaluating the nature and extent of the impact on the economy in general and the living standards in particular, of the realizations of various scenarios concerning the availability and the mix of raw energy and the type of conversion technology utilized. However, simple, rough calculation below shows that any such model can become quite large and perhaps unmanageable if sufficient care is not exercised in its development.

The input-output matrix provides a convenient vehicle for incorporating into a mathematical programming model the technological and many of the economic interactions of the economy. Despite its shortcomings, such as constant returns to scale, fixed technology and time delays involved in data collection and publication, it is attractive because it provides an internally consistent and a single most comprehensive data source. In its standard published form, it is available as an 87-sector matrix.[1] Next, the energy sector may be modeled by approximately 150 equations per period including the capacity constraints on activity levels of the energy processes and interperiod capacity carryover constraints. See [12], for example. An order of magnitude for the number of constraints per period in an integrated model with a reasonable level of detail is therefore computed to be 400: 87 for industrial activity, 2×87 for constraints of capacities on levels of industrial activity and for capacity carryover from one period to the next, and about 150 for the detailed energy sector. A 20-25 period model (e.g. a 25-year annual model, or a 75-year triannual model, etc.) would therefore have approximately 8000-10,000 constraints and much more if more detailed input-output matrix and energy sectors are employed. While linear programming models of this magnitude are certainly not considered to be impossible to solve, they would be among the largest models built to date. More importantly, preparation, testing/validation, and production runs for such a model would most likely consume both substantial sums of money and substantial amounts of time.

The aim of this exercise is not to suggest that one build such large models but rather to draw attention to the potential model size resulting from indiscriminate modeling and the difficulties that may arise.

It is therefore absolutely essential that a critical and scientific assessment be made of the exact nature of the formulation, scope and limitations of such a class of models. In particular, it is important to obtain answers to questions along the following lines:

---

[1] A much more detailed 367-sector matrix is also available from the Department of Commerce.

Formulation--In specific terms, what aspects should be modeled (endogenous), and what aspects should be assumed (exogenous) and what information should flow between periods? What linkages between the energy sector and the economy should be formulated and how?

Availability of data--What are the data requirements of the model? Are such data available? If not, is it possible to obtain satisfactory quick-and-dirty estimates to satisfy the immediate needs? And, what types of studies are needed to develop better quality data over a longer term?

Information from the model--What type of meaningful information can the model provide? What are the different objective functions that can be evaluated? At what level of detail should the model be formulated to provide the information desired?

Computation of solutions--Can the model be (efficiently) solved on the computer? What would the computational costs be? What refinements or special purpose algorithms exist (that perhaps require further research and) that can substantially reduce the computational costs?

These and other similar considerations point towards a need for developing and experimenting with a much smaller model that incorporates many, if not all, of the essential features of its larger counterpart. Our pilot model is an attempt to satisfy such a need. We believe that it will be small enough so that when implemented on the computer, it will have the agility for extensive experimentation. On the other hand, we also expect that it will incorporate most recent available data of sufficiently good quality (albeit in an aggregated form) that it can also be used to generate some meaningful scenarios showing how the economy might be affected if the energy picture evolves in a specified way.

In what follows, we first describe the model in some detail. Next, we give a brief and general mathematical statement of the model. Finally, we briefly review the current status of the model, its mathematical structure, and possible solution approaches.

2. DESCRIPTION OF THE MODEL. In the model, a 23-sector input-output matrix represents various industrial processes of the economy (Exhibit 1). The net output from the industry, together with net imports, meets the national bill of goods for consumption, capital formation and government services. The energy demands of the economy are met by the activities of the energy sector. The nature and extent of the capacity expansion in both the energy sector and the rest of the economy are endogenously determined. Finally, the exogenously given workforce provides the manpower necessary to sustain industrial production, energy processing and capacity expansion.

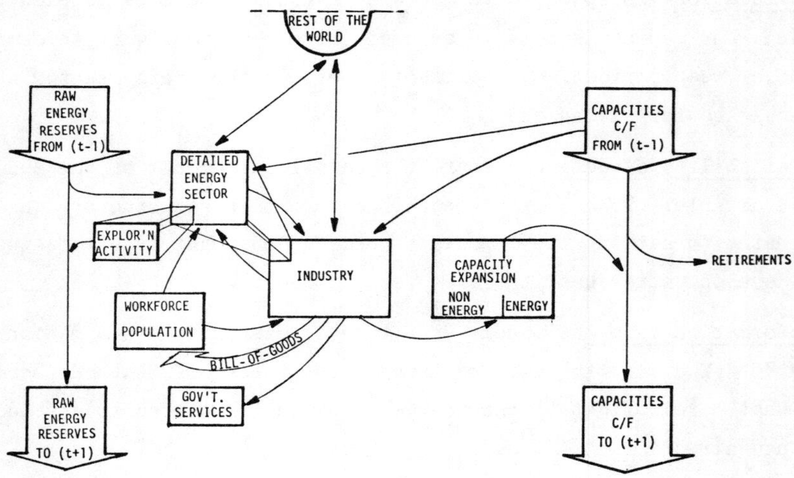

EXHIBIT 1. A SCHEMATIC REPRESENTATION OF THE MAIN LINKAGES IN THE PILOT MODEL

The detailed energy sector in the model includes technological description of the raw material extraction and energy conversion processes (Exhibits 2 and 3). Uranium mining, milling, conversion, enrichment and fabrication, light water reactor, fast breeder reactor, and spent fuel reprocessor are among the nuclear fuel based processes in the model. Oil and gas exploration and production, oil refining, gas transmission, coal mining, power generation using coal, oil and gas, and coal gasification and liquefaction are among the fossil fuel based processes in the model. The operating levels of the processing units are limited in one way or the other by the available capacities and proven reserves in any period. The proven reserves may be augmented by the exploration activity. And, raw material imports/exports make up the difference between the domestic production and usage.

Among the linkages that interconnect the energy sector and the rest of the economy are (Exhibit 1): energy demands of the economy, total manpower available to all sectors (including energy) of the economy, favorable balance-of-payments requirement, and bill-of-goods needed for energy processing and capacity expansion.

In order to mitigate many of the distortions caused by price changes and inflation, the industrial processes of the National Economy and the detailed energy sector will be represented in terms of physical flows. For the energy sector this is relatively easy to do because its activity can be treated in BTU terms. For the non-energy sector however, it is more difficult because (i) most industries produce a heterogeneous product thereby creating a need for developing a weighted index of the component physical outputs, and (ii) the input-output transactions are compiled in dollar terms, and money quantities depend on prices as well as physical flows. Moreover, the component prices unfortunately vary relative to one another over time, and so do the relative magnitudes of the component outputs. If these relative price and output variations among the components are assumed to be absent,[2] then a weighted index can be conveniently obtained by defining a composite product for the heterogeneous industry using base-year prices as weights for base-year outputs. The dollar transactions are thence reinterpreted as physical units of the composite product.

Whereas the input-output matrix represents the operating coefficients of the industrial processes of the economy, the capital coefficient matrix represents the amounts of the industrial products needed for a unit of (output) capacity expansion in any industrial (or energy) sector. Because a portion of the available capacity is retired at the end of each period, some capacity addition would be required in any scenario just to sustain the capacity of a process at a fixed level. This feature of the model also makes possible the process substitution. Thus, in the detailed energy sector where we expect to incorporate data of the new energy conversion technologies such as coal gasification and liquefaction, fuel cells, fast breeder reactors, etc., it will be possible to examine scenarios in which the distribution of capacities across the energy conversion processes evolves over time to reflect the impact of a particular set of assumptions specific to a scenario. On the other hand, in order to avoid the effort and difficulties involved in compiling reliable data of a similar nature for each of the other sectors of the economy, these will be initially represented by a nonvarying input-output matrix without substitution.

One of the primary linkages between the economy and the detailed energy sector is that of the energy sector meeting the demands of the economy. These energy demands are made up of the following four components: energy required for industrial processing, energy for personal (family) consumption, net exports of processed energy, and energy required to provide government services. In the model, these demands are transmitted to the energy sector in terms of the following four final energy forms:[3] oil products, gas products, coal and electricity. Moreover, this same set of demand variables is employed to compute the amounts of industrial goods and services required for energy extraction

---

[2]Unless specific allowance is made, this assumption would be implicit in any temporal input-output model.

[3]An alternative level of information detail would consist of eight final energy forms using the data developed by Knecht and Bullard [1975]. We may experiment with this form of the linkage at a later date.

and processing.[4] The latter linkage also requires a modification of the input-output matrix.

The activities of the detailed energy sector are represented in two groups: nuclear and non-nuclear. The non-nuclear group contains for the most part the fossil fuel based activities. It also includes the fuel free activities such as hydroelectric, geothermal, etc.

Exhibit 2 schematically shows the electric power generation related activities of the nuclear fuel cycle in the model. Natural or recycle uranium goes through chemical conversion and physical separation and enrichment before it is fabricated into the fuel elements for the light water reactor (LWR). Fuel elements could also be fabricated from recycle plutonium and uranium. The spent fuel may be reprocessed to recover the plutonium and uranium. The liquid metal fast breeder reactor (LMFBR) operation is similarly defined in the model.

Exhibit 3 shows the activities of the fossil fuel based energy processes in the model. Exploration for either oil or gas results in additions to the reserves of these raw energy forms. Oil and gas production, and coal mining activities provide the raw fossil fuels which are next processed into final energy forms. For oil, this involves a refining activity that produces some oil products (gasoline, heating oil, etc.) for satisfying final demands, and other oil products (residual fuel oil) for use in electric power generation. Due to nature of the linkage by which the energy sector meets the energy demands of the economy, the detailed yield structure of the refinery operations is not represented here. Natural gas is transmitted either to meet the final demands or for power generation. For coal, three alternative uses are defined in the model: to meet coal demands of the economy, for power generation, and for synthetic oil and gas.

One of the most important linkages in the model requires that all capacity building be constrained by the capacity of the economy to build capacity, either for capacity expansion or for replacement of the retired equipment. This set of constraints imparts a tendency in the model for a <u>gradual</u> evolution over time of the distribution of the capacities across the exploration, production, and conversion processes of the energy sector. Because, if too drastic a change in the capacity distribution were to occur, it would most likely drain an unusual amount of the economy's capability to build capacity and leave an insufficient ability to build capacity of the other industrial processes. On the other hand, the economy could expand at an unusually rapid pace its capacity expansion

---

[4] Using data similar to those developed by Just et al. [1975], it is possible to incorporate a full blown operating coefficient matrix to more accurately provide this linkage.

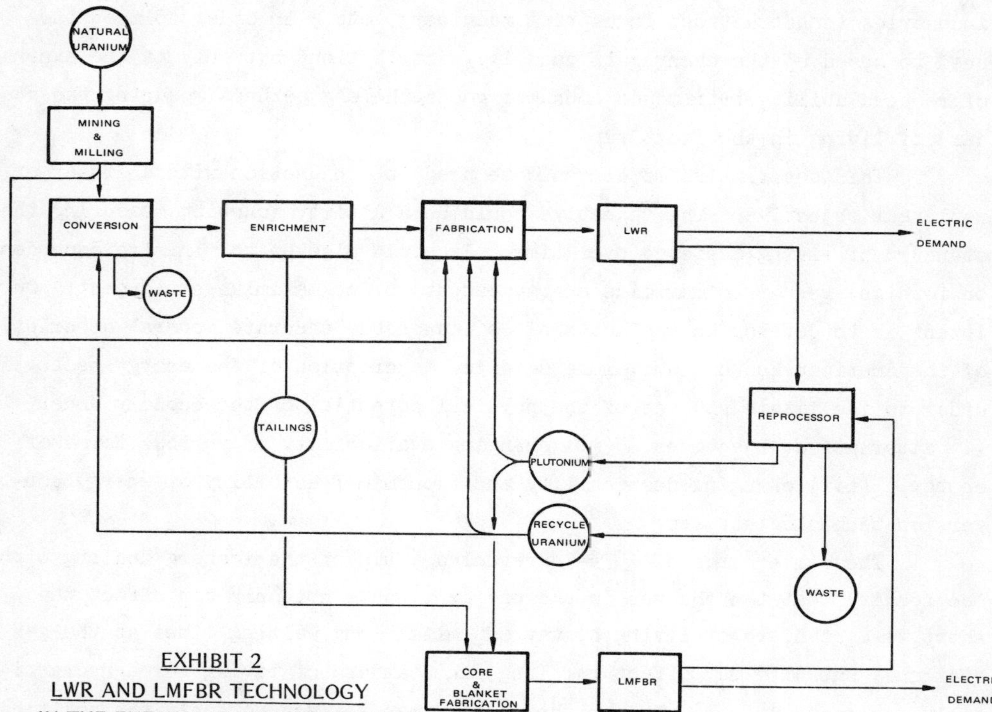

EXHIBIT 2
LWR AND LMFBR TECHNOLOGY
IN THE ENERGY SECTOR OF PILOT
(SUGGESTED BY: T. J. CONNOLLY), 4/75

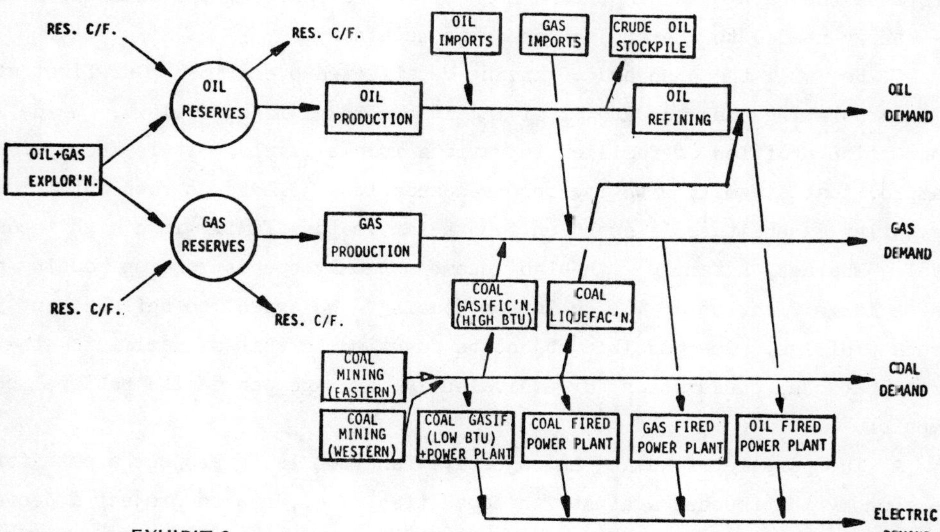

EXHIBIT 3
FOSSIL FUEL BASED ACTIVITIES
OF THE ENERGY SECTOR OF PILOT

(PRELIMINARY- 6/75)

industries (construction, industrial machinery, etc.) in order to meet the need to speed up the changes in capacity distribution, but only at the expense of reduced ability to produce consumer goods thereby perhaps reducing the standard of living in the short run.

This descriptive model could be used in conjunction with a linear or a nonlinear objective. The objective could be a utility function measuring the standard of living achieved over time. It could also be to minimize dependence on foreign ore, or to maximize energy output, or to maximize employment. Our intent is to develop on a pilot scale a reasonably accurate general description of the American Economy and a more detailed description of the energy sector in order to facilitate studies of the physical potential of the economy under (i) alternative objectives, (ii) changing availability of various forms of energy, (iii) changing desirability and economic feasibility of energy conversion technologies, etc.

The oil embargo of 1973-74 revealed a tip of the iceberg dealing with the reality that the changes in the energy picture not only can affect the short term standard of living by way of (mile) long waiting lines at the gas stations, but also may affect the long run standard of living through drastically increased prices. Such higher prices may reflect not only the political realities of the world's raw energy markets but also much increased physical effort on the part of the American Economy to provide from the domestic sources the energy needed to operate the economic machinery.

How will the standard of living be affected over time? Our first stab at incorporating the standard of living in the model is as follows. We define consumption profiles of families in various income levels. It is known, for example, that a family with low income spends less dollars on food but more of its dollar expenditure is spent on food relative to a family in a high income level. Whereas, a family with high income not only spends more on housing but also a large fraction of its dollar on housing. We expect to define about 5 to 7 such profiles. One possible objective function is that of maximizing the "gross national consumption" or equivalently "average per capita national consumption."

The purpose of having an objective function is to project a path for the economy that pushes against its capacities, i.e., not to project a depression economy. In examining the question of the objective, however, one is immediately faced with the prospect of finding a generally acceptable utility (or welfare) function for the entire country--a not too promising task, to say the least. A much more plausible approach is to incorporate some information on national welfare in the objective function and some in the constraints. The maximization of gross national consumption as defined by the income level profiles is one possibility. In any case, one thing seems certain. It will require a great deal of experimentation before a satisfactory objective function is realized.

3. GENERAL MATHEMATICAL STATEMENT OF THE MODEL. In the model, there are interperiod and intraperiod constraints (Exhibit 4). They are briefly outlined below. A somewhat more detailed description can be found in [5].

The interperiod constraints connecting periods t and (t+1) appear below the lower dashed line in Exhibit 4. These are capacity balance constraints, manpower skill adjustment limit constraints and those related to raw energy reserves, cumulative exploration and production, and intermediate energy stocks. The capacity balance (or capacity c/f) constraints specify that the available capacity in period (t+1) of any activity equals its capacity in period t, less retirements plus capacity built. Next, the manpower is assumed to be made up of several skill groups, e.g., unskilled, skilled, engineers, managers, etc. The manpower skill adjustment limit constraints (manpower c/f) specify the educational and training limitations. This set of constraints, together with the intraperiod constraint that the sum over all skill groups cannot exceed available workforce, provides for changes in the size of skill groups to satisfy the manpower needs.

The following three sets of interperiod constraints are intended to keep track in detail of the energy reserves, cumulative exploration (and production), and stocks. The reserves constraints specify that

$$\begin{bmatrix}\text{Reserves in}\\ \text{period (t+1)}\end{bmatrix} = \begin{bmatrix}\text{Reserves in}\\ \text{period t}\end{bmatrix} - \begin{bmatrix}\text{Raw energy}\\ \text{extracted}\\ \text{in period t}\end{bmatrix} + \begin{bmatrix}\text{Additions to}\\ \text{reserves in}\\ \text{period t}\end{bmatrix}$$

Cumulative exploration in say, feet drilled (and production in say, BTU's extracted) is determined as follows:

$$\begin{bmatrix}\text{Cumulative exploration}\\ \text{at the beginning of}\\ \text{period (t+1)}\end{bmatrix} = \begin{bmatrix}\text{Cumulative exploration}\\ \text{at the beginning of}\\ \text{period t}\end{bmatrix} + \begin{bmatrix}\text{Exploration}\\ \text{during}\\ \text{period t}\end{bmatrix}$$

The production of an energy form may be direct as in the case of oil, gas, uranium, etc., or a byproduct of some other activity, as in the case of plutonium. Finally, the stock (inventory) balance constraints: For ith energy form,

$$\begin{bmatrix}\text{Stocks at the}\\ \text{beginning of}\\ \text{period (t+1)}\end{bmatrix} = \begin{bmatrix}\text{Stocks at the}\\ \text{beginning of}\\ \text{period t}\end{bmatrix} + \begin{bmatrix}\text{amount produced}\\ \text{during}\\ \text{period t}\end{bmatrix} - \begin{bmatrix}\text{amount used}\\ \text{during}\\ \text{period t}\end{bmatrix}$$

EXHIBIT 4

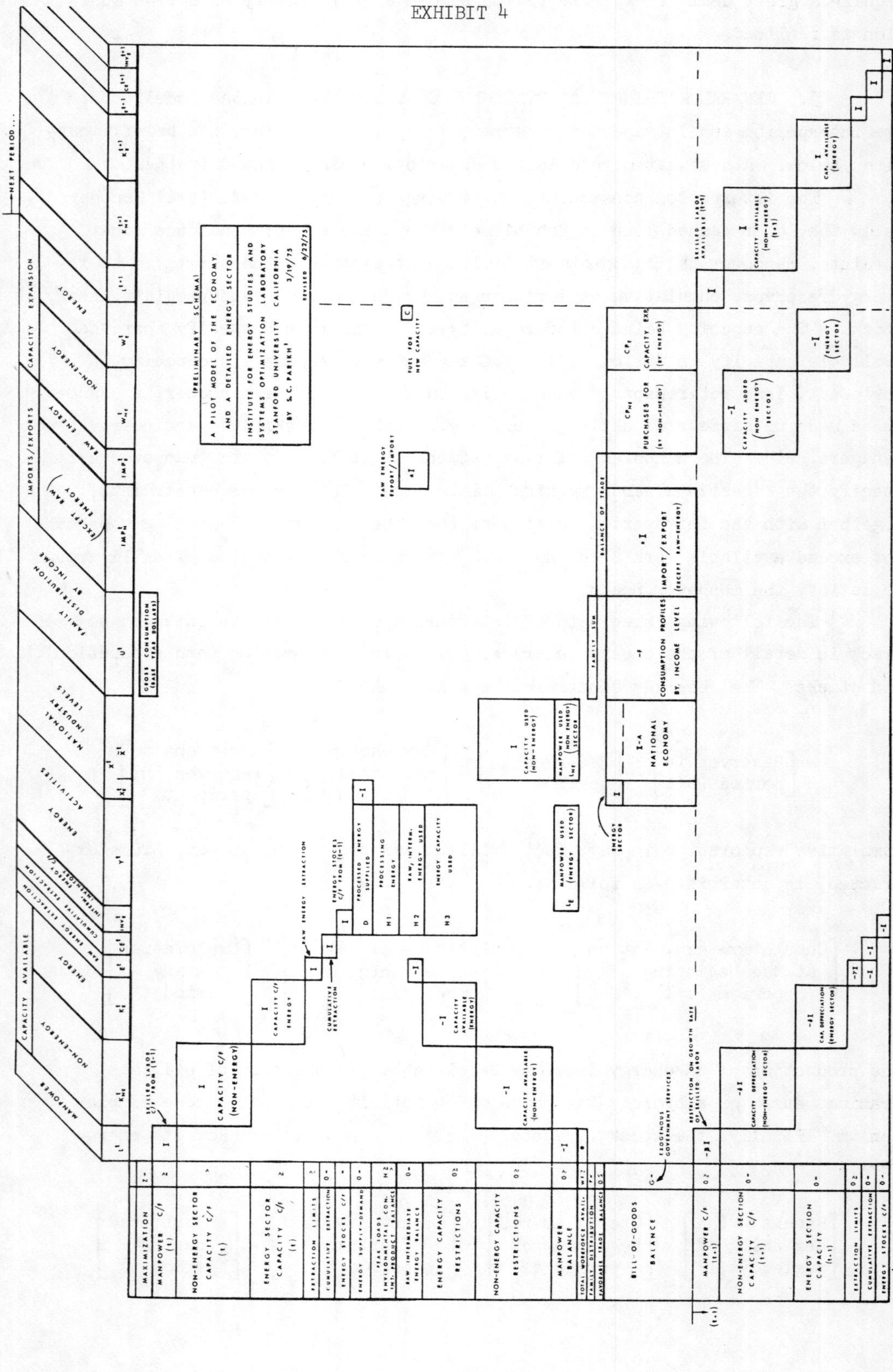
(PRELIMINARY SCHEMA)
A PILOT MODEL OF THE ECONOMY AND A DETAILED ENERGY SECTOR
INSTITUTE FOR ENERGY STUDIES AND SYSTEMS OPTIMIZATION LABORATORY STANFORD UNIVERSITY CALIFORNIA
BY S.C. PARIKH
3/19/75
REVISED 6/22/75
NEXT PERIOD...
MANPOWER
CAPACITY AVAILABLE
NON-ENERGY
ENERGY
NATIONAL INDUSTRY LEVELS
FAMILY DISTRIBUTION BY INCOME
IMPORTS/EXPORTS
CAPACITY EXPANSION
MAXIMIZATION
MANPOWER C/F
NON-ENERGY SECTOR CAPACITY C/F
ENERGY SECTOR CAPACITY C/F
ENERGY CAPACITY RESTRICTIONS
NON-ENERGY CAPACITY RESTRICTIONS
MANPOWER BALANCE
BILL-OF-GOODS BALANCE
NON-ENERGY SECTOR CAPACITY C/F
ENERGY SECTOR CAPACITY
CAPACITY C/F (NON-ENERGY)
GROSS CONSUMPTION (BASE YEAR DOLLARS)
NATIONAL ECONOMY
CONSUMPTION PROFILES BY INCOME LEVEL
FAMILY SUM
BALANCE OF TRADE
IMPORT/EXPORT (EXCEPT RAW-ENERGY)
PURCHASES FOR CAPACITY EXP.
RAW ENERGY EXPORT/IMPORT
FUEL FOR NEW CAPACITY
CAPACITY AVAILABLE (NON-ENERGY)
CAPACITY USED (NON-ENERGY)
MANPOWER USED (ENERGY SECTOR)
ENERGY SECTOR
UNIVERSITY OF SANTA CLARA, CALIFORNIA
Drafted by Randall Chang

Whether a constraint from these 3 sets is included in the model for a particular energy form depends upon its need and/or validity. For example, one may leave out the stock balance constraint for natural gas by arguing that gas could be extracted only if needed during the period. On the other hand, the exploration constraint for plutonium is invalid. The constraints from these sets in Exhibit 4 are for illustration only.

The intraperiod constraints appear between the two horizontal dashed lines in Exhibit 4. The first set (involving matrix block D) provides for meeting the energy demands of the economy from the energy sector. The next two sets (involving matrix blocks H1 and H2) represent various energy processing aspects. The environmental aspects of energy extraction and conversion could also be included here. The next two sets (involving matrix block H3, and variables x^t, K_{NE}^t) specify the operating capacity limitations of the energy and non-energy processes. The next set (involving L_{NE} and L_E) specify the manpower constraints that for each skill group, the manpower used cannot exceed that available. The next constraint states that manpower sum across skills cannot exceed available workforce. The family sum equation is used in conjunction with the objective function described below. The balance of trade equation computes the trade balance in each period for the purpose of incorporating a favorable trade balance requirement. Such a requirement may be imposed individually in each period or collectively in several periods. Finally, the bill-of-goods balance equations specify that the industrial output, together with imports (IMP) meets final demands consisting of personal consumption (Fu), exports (EXP), capacity expansion ($CP_{NE}W_{NE} + CP_EW_E$), and government expenditures (G).

Now consider the form of the objective function in the model. Broadly speaking, the objective of the model is to maximize the discounted vector bill-of-goods received per person summed over time. Suppose that in the base year, the physical bill-of-goods for people with consumption level M_k (income less taxes and savings in base year dollars) is: $b^k = [b_{1k}, b_{2k}, \dots, b_{nk}]^T$, $k = 1,\dots,K$ and with $M_1 < M_2 < \cdots < M_K$. Let u_k^t be the unknown number of people in period t that receive b^k. Then, $u_1^t + u_2^t + \cdots + u_K^t = P(t)$, the population at time t. The total bill-of-goods for period t is Fu^t, where: $F = [b^1, b^2, \dots, b^K]$, and $u^t = [u_1^t, u_2^t, \dots, u_K^t]^T$. Initially, the overall objective will be to maximize discounted gross national consumption over time, i.e., maximize $\Sigma_{t=1}^T \lambda_t \, GNC(t)$, where $GNC(t) = M_1u_1^t + \cdots + M_Ku_K^t$, and λ_t = weight in period t for discounting. It should be noted here with caution that the treatment of the objective function may change, even drastically, as experiments are performed on this model and numerical results become available. For a discussion of the use of production functions, demand functions and other forms of the objective, see [4] [4a].

Finally, unless specific allowance is made, an optimal solution to the model may turn out to be such that all (or most) capacity is depleted by the end of the time horizon T. Such unrealistic end effects can be avoided in several ways. One is to put a much higher weight λ_t on GNC(T), the gross national consumption in the last period. Such a weight conceptually would reflect the present value of consumption beyond T. Another way is to specify the terminal capacities generated by an equilibrium model or a steady growth model. For the energy sector, this specification could be in gross BTU terms across several processes, thereby allowing for changes in capacity distribution across processes.

4. FURTHER REMARKS. The model formulation and data source identification are almost complete. Presently, the data are being aggregated and a model is being prepared for computer solution. We expect the model to have about 125 equations per period. For a 30-year triannual model, there will be about 1250-1400 equations, including the specification of initial capacities and end effects. Initially, the model will be solved using the straight simplex method of the MPS/370 system.

In order to pave a way for economical solution of similar much larger problems having 8,000-10,000 rows referred to earlier, it is also expected that the PILOT model will provide us at the Systems Optimization Laboratory a prototype for research in solving large-scale linear programming models of energy systems.

The PILOT model belongs to a class of models having a staircase structure (Exhibit 5) that often arises in dynamic linear programs. For such time-phased problems, the number of iterations to optimum may be as high as 10 times the number of rows as opposed to widely experienced 2 to 4 times the row count in unstructured problems [Beale, 1971].

Several special purpose algorithms are available that take advantage of the staircase structure for efficient solution. Computational results on some of the methods show that this is a partly-proven and a very promising research area. See, for example Dantzig [1963, Ch. 23], Glassey [1973] and Ho [1974], just to mention a few.

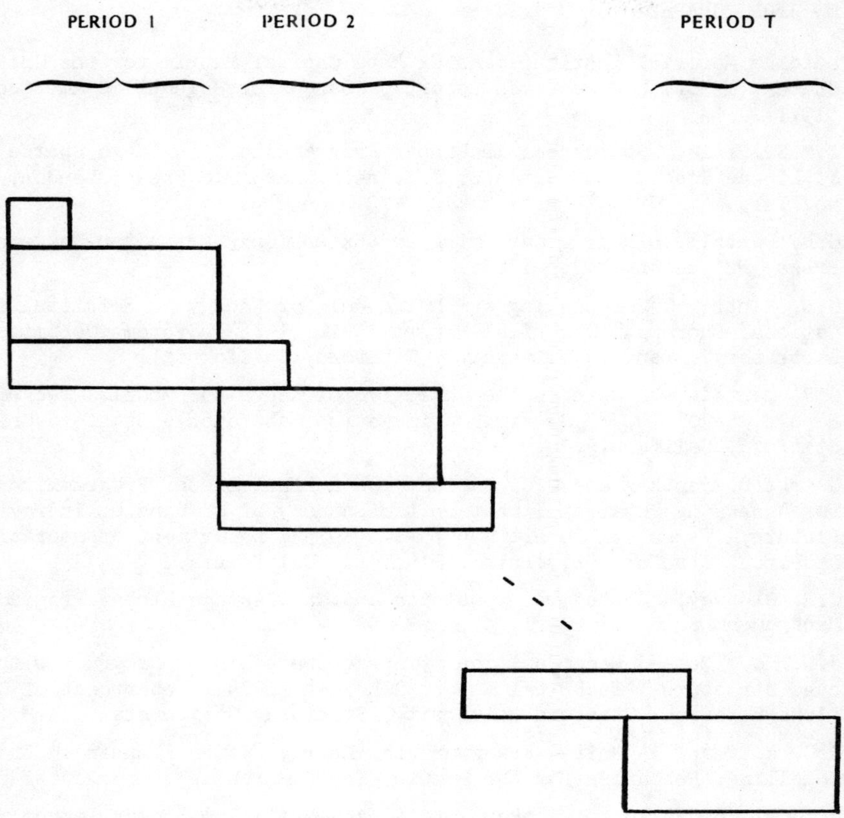

EXHIBIT 5. THE STAIRCASE STRUCTURE OF PILOT

SELECTED BIBLIOGRAPHY

1. Batelle Memorial Institute, An Ex Ante Capital Matrix for the United States, 1970-75, A research report prepared for Scientific American (1971).

2. E.M.L. Beale, "Sparseness in Linear Programming," in Large Sparse Sets of Linear Equations, edited by J.K. Reid, Academic Press, London, 1971, pp. 1-15.

3. G.B. Dantzig, Linear Programming and Extensions, Princeton University Press, Princeton, N.J., 1963.

4. G.B. Dantzig, "Formulating a PILOT model for Energy in Relation to the National Economy," Technical Report 75-10 (1975), Systems Optimization Laboratory, Stanford University, Stanford, California.

4a. G.B. Dantzig, "A Note on the Objective of the PILOT Model," Technical Report 75-20 (1975), Systems Optimization Laboratory, Stanford University, Stanford, California.

5. George B. Dantzig and S.C. Parikh, "On a PILOT Linear Programming Model for Assessing Physical Impact on the Economy of a Changing Energy Picture," Technical Report SOL 75-14 (1975), Department of Operations Research, Stanford University, Stanford, California.

6. C.R. Glassey, "Nested Decomposition and Multi-stage Linear Programs," Management Science, 20 (1973), 282-292.

7. J.L. Ho, "Nested Decomposition of Large Scale Linear Programs with Staircase Structure," Technical Report SOL 74-4 (1974), Department of Operations Research, Stanford University, Stanford, California.

8. K.C. Hoffman, "A United Framework for Energy System Planning," Energy Modelling, Resources for the Future, Inc., March 1973, pp. 108-143.

9. J. Just, B. Borko, W. Parker, and A. Ashmore, "New Energy Technology Coefficients and Dynamic Energy Models," Vols. I and II, Mitre Corporation Report MTR-6810, January 1975.

10. R.L. Knecht and C.W. Bullard, III, "End Uses of Energy in the U.S. Economy," Center for Advanced Computation Document No. 145, University of Illinois at Urbana-Champaign, January 1975.

11. A.S. Manne, "U.S. Options for a Transition From Oil and Gas to Synthetic Fuels," Kennedy School of Government, Harvard University, Cambridge, Mass., December 1974.

12. W. Marcuse, L. Bodin, E. Cherniavsky, and Y. Sanborn, "A Dynamic Time Dependent Model for the Analysis of Alternative Energy Policies," Brookhaven National Laboratory, Upton, N.Y., 1974.

Stanford University
Stanford, California 94305
and
University of Santa Clara,
Santa Clara, California 95053

Proceedings of Symposia in Applied Mathematics
Volume 21
1977

THE PROBLEM OF AGGREGATION IN MODELING PHYSICAL AND SOCIAL SYSTEMS AND PROCESSES

Richard L. Garwin

ABSTRACT: The design of complicated systems (nuclear reactors, bridges, advanced electronic systems, business organizations) usually proceeds on the assumptions of restricted interaction, local linearity, and similar simplifying approaches. Once such a system is designed it may be analyzed further by removing the simplifying assumptions, often one at a time. But "really to know" what goes on in a postulated nuclear-reactor accident; understanding the behavior of society following a major perturbation like the Arab OPEC embargo; judging in general the behavior of a complex system far from the design point; requires for efficient simulation some degree of aggregation in the treatment of the system. Insufficient aggregation poses problems of computer capacity; excessive aggregation may lose important aspects of behavior, as in some energy-related examples discussed. Furthermore, societal models must allow interaction with people motivated to act the roles of cartel operators, anarchists, law-breakers, and the like. To exclude from the model sectors of society which are much in demand, such as lawyers and lobbyists, is to neglect an important influence.

This talk prescribes no solution to the stated problem nor even provides well-formulated questions, but, I hope, rather indicates the importance and opportunity in this field.

How is it, with all the mathematical, scientific, and engineering talent in the United States; with our capital and productive resources; that we have an "energy problem"? Will whatever in the past limited our success in this field do so in the future? Can any finite set of reforms make a difference?

AMS(MOS) subject classifications (1970). Primary 90A00, 90A15, 90B99.

BACKGROUND:

I come to this question from some 25 years of contact with technical questions of increasing involvement with national policy. In the summer of 1950, I organized a group at the Los Alamos Scientific Laboratory and began an experiment to remeasure the reaction cross section of deuterium on tritium and of deuterium on deuterium to help decide whether it would be feasible to make thermonuclear weapons or fusion energy. From 1951 to 1954 I worked part time at Los Alamos on the hydrogen bomb and became acquainted with approaches to fusion energy, the problems of fallout and the use of computers in modeling man-made or natural systems.

From 1962 to 1965 and 1969 to 1972 I was a member of the President's Science Advisory Committee (PSAC). For the PSAC I was Chairman for a long time of the Antisubmarine Warfare and Military Aircraft Panels. That latter developed into an Aircraft Panel concerned with civil transportation, and even an Air Traffic Control Panel. These and the Defense Department activities led to acquaintance with operations research and with the political system whereby decisions were taken. One of the interests of the Aircraft Panel over some years was the Department of Transportation Study of Transportation in the Northeast Corridor. As a member of PSAC, I was also quite concerned with the mid-1960's massive energy study of the Office of Science and Technology, led by Ali Cambel.

From 1970 to 1974, following the Clean Air Amendments of 1970, I was a member of the National Academy of Sciences Committee on Motor Vehicle Emissions which was requested by the Congress to study the technological feasibility, economics, and possible necessity of delay in reaching Congressionally mandated emission standards from automobiles. However, our charter did not extend to a determination of the "best way" in our society (or a better way) to accomplish the Clean Air Amendment goals of air quality, even in the automobile sector, as by an emission tax instead of absolute limits on motor vehicle emissions. It was clear to some of us that there were alternative incentive structures which could accomplish the same goals more certainly and which would inherently provide better motivation for the automobile manufacturers.

In December 1973, I chaired a Task Force for the IBM Research Division to see what our division, our company, and the nation might do to mitigate the effects of the oil embargo and of higher fuel prices.

In 1974 to 1975, I participated in the American Physical Society Light Water Reactor Safety Study and became acquainted with the complicated conditions of flow which must be taken into account in the analysis of potential reactor accidents. In 1975 I chaired a National Academy of Sciences Study on the Establishment of a Solar Energy Research Institute by

ERDA. At present, I am a member of the Federal Energy Administration Environmental Advisory Committee and of a Ford-Foundation-sponsored Nuclear Energy Policy Study.

In 1965 as a result of the Northeast Power Blackout I helped to begin IBM Research's study of large interconnected electric power systems.

In the mid-1960s I first had some contact with the US SST Program as Chairman of the Military Aircraft Panel and later in 1969 chaired a Special SST Review Panel of the Office of Science and Technology. A year or so later, I restudied the problem at the request of the relevant congressional committees and had a rerun this year in regard to the decision on landing rights for the Concorde aircraft. The problems now are those identified in 1969--airport noise, economics (especially fuel consumption), and technology, plus one more--the effect of exhaust gases in catalytically destroying the ozone layer.

We have problems. I think that if decision makers could see the future, they'd make better choices. And if the people who are watching the decision makers could see the future, then the decision makers would certainly make better choices. But where the future is murky, it's all too easy to choose one route or another because it sounds plausible or because it can be made to sound plausible. Even if the future cannot be seen in all detail, some candidate programs can be compared against others against some set of values. That would be useful. So we ask whether the future can be seen through computer modeling of the proposed systems. If not--if you say no--then how can we design systems with assurance? How can we choose a policy, whether a tax policy or military force structure, and have confidence in the outcome? If we can't model, how can we answer questions regarding reactor safety, worth of investment in fusion research, and so on?

People in the business have several approaches which are to some extent in competition. Some practitioners look down upon those who want just to simulate in fast time, using the computer as a kind of experimental prototype, which, started with assumed conditions and policies, will show more or less how things would be if these policies were in fact carried out. Many prefer to use some kind of automatic control algorithm in which one has not only some kind of model of the system, which might be a mathematical model in some closed form or a computer model, but in addition a procedure for restoring the system to some desirable state or for minimizing some loss function. But optimization without straight simulation seems quite inadequate, because of the need to show non-technical people--legislators, administrators, voters--how the <u>present</u> system is going to work or not work and how any alternative system will compare. So I think one is brought back to some kind of fast-time simulation even if one has a control algorithm, and one has to be able to handle people as well as machines in this simulation process.

EXAMPLES OF SYSTEMS PREDICTABLE BECAUSE CONSTRAINED:

Often in systems which are synthesized--artifically constructed--it is better to sacrifice some hypothetical efficiency in favor of assurance of costs of implementation, schedule, and reliability. For instance, take Personal Rapid Transit, defined here as having mostly four-passenger vehicles which traverse a city mostly on elevated guideways (this is all hypothetical--none of these exists). One has a problem in scheduling these vehicles, controlling them so that they don't collide, and in knowing and raising the saturation factor of the system (the fraction of the guideway that can be filled with vehicles moving at 40 or 60 miles an hour). In this case I think we ought to have initially a rather simple ("synchronous") control algorithm in which each of the vehicles normally follows a moving slot, a point in space which moves along at constant speed, and where the control restores the vehicle to that target slot. This is the analog of a lattice gas rather than a continuous mechanical system, and is obviously not an optimal control, but I think it's good enough. One needs bypass tracks for accelerating the vehicles and can bring them up to speed at an appropriate time to merge with an empty slot on the main track, and the same for transferring from one line to a line at right angles. I think one can get 60 or 70% occupancy by a quasi-synchronous control which limits long-range interactions by allowing limited transitions into adjacent slots. But what's more important, you could design and field such a system <u>now</u> so far as the control law is concerned, whereas experts in the field tell me there is no optimal control algorithm known. It will be several years before they can do what will satisfy them professionally, and I think that optimal control could be introduced later. So I think simulation of such a system (and of such an approach to system implementation) would show the merits of the approach I advocate, which is to constrain the design to make it controllable, to constrain the system to make it computable, and thereby sooner to realize the benefits of operation.

Similarly in air traffic control we published a report of the Air Traffic Control Panel of the President's Science Advisory Committee in 1971 (actually appeared in 1974) and proposed that one run the system with four-dimensional on-board navigation, where given a take-off time and a landing time each aircraft should follow a path in space as a function of time in order to show up to be handled routinely at its destination. So there would be very little real-time control required of the supervisory system; aircraft would handle their own navigation with respect to moving target points. In choosing such an approach one sacrifices. An aircraft cannot at all times make its maximum air speed because it has to track a point in space as a function of time, and so if adverse winds are felt the aircraft has to have some speed in reserve. Once again we sacrifice what might possibly be an improved system (slightly higher air speeds) for a system which is predictable and controllable. In this case the role of people (air traffic controllers and pilots) could be designed for efficiency and reliability. But overall social systems are different.

We reduce perhaps the ultimate efficiency of some social systems by reducing the information available, but may intervene less than in alternative societies. In the business sphere we properly forbid exchange of information among competitors. They set their prices independently, and that has been judged necessary if we are to maximize societal benefits and freedom. In other societies nominal competitors can exchange information and nobody cares because the price is set by the government. You have to choose one system or the other. You can't let suppliers pretend to have independent prices which are in fact not independent due to exchange of information.

The Northeast power blackout in 1965 involved an example of a large interconnected electrical power system. Such a system has tens of electrical generators, transmission systems, and thousands or hundreds of thousands of motors and other loads. Each of these generators or loads can be analyzed readily if it's working against some rigid fixed system, a given voltage at its terminals and near some design point. But how far this analysis, how far from stability the whole system was, was brought very clearly to our attention in 1965 when in New York City it was discovered experimentally that hospitals didn't have emergency lights; that La Guardia Airport carefully drew its power from three individual substations, all three of which were on the same electrical power grid and so the airport went black as airplanes were landing; that major generators in New York City had oil pumps for their bearings which were driven from the electric lines so that when the generator was coasting (grinding) to a halt because there was no electricity to run the pump, it couldn't resume operation for several months.

It's clear that a full analysis needs a lot of details, and that's really the subject of my talk--the degree of aggregation that one can use in describing the important aspects of a system. The level of detail goes to the moment of inertia of the rotating systems, the design of the governors and the steam throttles which drive the turbines. One might choose to require a "good citizen"--a kind of generator which is very stable and easily controllable when put into an electrical power net. And in doing this one sacrifices something overall--another example of increasing the margin of stability by sacrificing some potential efficiency. Since 1965 one has done other things; e.g., introduced load shedding and better communications in order to be able to keep systems from collapsing. What one needs to do first before one can control a system is to understand the consequences of the assumed systems. Then to improve the simulation and to improve the control really is inexpensive for these large systems relative to the benefits that can be obtained, but it may take time. For instance, if electric power systems run with about 20% spinning reserve, they have 20% more generating capacity on-line than is required to handle the load. If one has a load shedding system whereby there are customers who have agreed and who are not going to be damaged by shutting off their loads, then one can save spinning reserve and reduce capital investment in electric power system by a comparable amount, which could be worth tens of billions of dollars.

DETAILS ARE NECESSARY TO STRATEGIC PLANNING:

Take another and more current example which the Federal Energy Administration and its Environmental Advisory Committee have been assessing recently--to consider reducing the investment per unit energy consumed as electricity by spreading the use of electrical energy through the day. Several approaches include peak time of day pricing and peak load pricing. The task is to set a policy and create systems which will accomplish this result.

Let's look a bit more in detail what's involved with time-of-day pricing and spreading the energy load. This is a particularly interesting example, because it is likely to spread rapidly over the nation from those few locations where it has just begun to be practiced. Most important is to characterize the consumer. Hardly anybody knows what the householder or the commercial organization will do, even with present equipment, if there were a rate schedule such that electricity used in peak load periods costs much more than if off-peak. There are some loads with which we have experience, mostly electric water heating and those are usually run on timers. But in order to know how much benefit there is in this approach you have to characterize the consumer before and after education. If you just ask somebody on the street how much he would reduce his peak consumption if he were given a time-of-day meter he couldn't tell you. And probably wouldn't be able to tell you even after a short course in how good it is for society and how much it might reduce his costs. Having characterized the consumer, which requires some effort, one ought to assume a rate structure and then a tentative model should determine the consumption, the investment required, the profit, to see who benefits and how to apportion the benefit to the various sectors. And then one ought to recycle the rates, the kind of education, and decide whether this is a promising approach. Then one ought to do experiments with consumers to refine the structure and parameters of the model. How much would it cost for meters which will do peak-hour metering, provide remote operation of disposable load, and allow instantaneous pricing for optional loads? A customer who would like to use some equipment at a certain time of day could benefit from a meter telling him how much it's going to cost. And it might be much more expensive to dry the clothes at 4:00 p.m. than at 7:00 p.m. (or than at 4:00 p.m. on some particular and unpredictable day) especially if one is dealing with a nuclear power system where base load electrical energy comes reasonably cheaply but peak load is expensive because capital cost is high.

Who needs this kind of simulation? Utilities, Public Service Commissions, the Federal Energy Agency, manufacturers of all this equipment, the Congress. In fact, the people who make the policy and decide the investment structure need it most and earliest. But you don't have those answers until you do the detailed simulation, and a poor decision as to aggregation in this modeling may seriously impair the value of simulation.

SOME EXAMPLES OF CARE IN AGGREGATION:

There are simple examples of systems which are different if one aggregates supposedly identical elements of the system. One spring and one mass result in one resonant frequency. Two springs and two masses (identical), uncoupled, still have nominally one resonant frequency, but if there is any coupling there are two resonant frequencies. There is a splitting of the degeneracy into two eigenmodes which have different symmetry. This gives the possibility for instance of damping the antisymmetric mode without introducing damping of the symmetric mode and that's a way in which one could in fact in a physical system aggregate, that is put in more units of mass and spring without perhaps introducing instability. But there are different ways of stabilizing systems. For instance, if we take the example of two identical generators, which might have some instability of this kind when feeding a single load, one might introduce small signal stability with linear damping elements which are always in place, interconnecting the generators. If the generators get slightly out of phase currents flow. But these small signal elements, called upon to handle only rather small currents and voltages, might be overloaded if there were a large impulse. If one of the generators stopped one could burn out this damping element. The system could fail. So a problem in designing such a system is to understand the cost of this damping--of the thing which allows you to treat two generators as one, or two masses as one. As an alternative, one might allow the small signal instability, replacing this coupling at small signals by a threshold stabilizer which sees when the generators or the masses are sufficiently far out of phase and which can handle large currents easily because it's not doing anything most of the time. The dissipation of the two approaches is different; the cost is different.

This problem of aggregation arises even in simpler systems. Suppose you have a gas producer and a gas consumer, with consumption not perfectly matched to the production. Storage is indicated. So you put in a balloon. You can imagine a balloon which fills during the warm weather and contracts during the cold weather because of the gas being used to heat houses. The balloon is a perfectly reasonable thing. It may even save some money. Suppose now you have more producers and more consumers and want to double the storage capacity; it's a natural thing to put two balloons on that same pipe. But the result may not be what you expected. Two balloons next to one another do not grow and shrink at the same time. In fact when the two balloons are used one shrinks to nothing and the other has all the gas in it. That happens not only with rubber balloons but with soap bubbles, and there are several ways to avoid it. One could have a mechanism controlling inflow or efflux from the balloons so as to feed them equally; one could have mechanical coupling between them; or one could allow one balloon to fill and then use the other one, but means would be needed to keep that one balloon from overfilling. Thus although no problem is introduced by the aggregation of treatment of the gas molecules, balloon-type reservoirs cannot be treated in the same manner.

We want time-of-day pricing for electrical energy in order to have some social goal achieved by the individual actions of the consumers. If there were only one consumer, it would be easy for him to assess the consequences of his using electrical energy whenever he wanted it. He would see it was cheaper for him to spread his usage because he is the only one who pays for the investment in electrical generating capacity. He would make those adjustments to his consumption, in order to obtain the social benefit, because for a single consumer by definition the social benefit is all going to be his anyhow. Individual consumers of a public may see the facts. They may understand the theory. They may see the instantaneous load. But they may still reckon a short-term advantage over other consumers and even a relative long-term advantage by not doing what we would define as the socially desirable course. And that's why we need the signals that come from a price system which reflects total costs. Perhaps formerly religion in a rather homogeneous society induced individuals to behave in a way so as to maximize a particular overall social benefit. But Gresham's Law at times takes precedence over religious law, and people with identical interests sometimes see that their local interest, local in time and space, is more toward taking from their colleagues than in maximizing the overall social good.

ENERGY POLICY AND THE NEED FOR SIMULATION:

To consider aggregation in reactor safety and how one might perhaps modify reactor design in some ways to make them easier to compute and therefore easier to certify as safe would exceed the time available. But I would like to discuss overall energy policy. In the middle sixties it was perfectly clear that one had options of quotas, tariffs, some choice of methods to limit or to increase US dependence on foreign oil or US consumption of foreign oil. The economic incentive was for this country to use low-cost oil from Saudi Arabia. It comes out of the ground at a cost of 15 cents a barrel, I understand, rather than the higher cost of US oil, which may cost \$2-\$4 per barrel to produce. In fact as one looked decade-by-decade into the future both the economic and the strategic incentive was to exhaust foreign oil first so that we would have our own oil left. On the other hand, also from a narrow US point of view, we wanted to avoid dependence on foreign oil so that in wartime or in case of other embargo we would have fuel to run factories, cars, and the like; so that we would have a current energy source. At that time my solution to this problem was to do more extensive drilling in this country, so that we would have enough production <u>capacity</u> to supply the instantaneous demand. These wells should then be produced at a low rate so that they would last for a very long time. My proposal was to have adequate refining capacity in the US to handle all US consumption. This would provide strategic reserve in the ground, would provide the production capacity, and thus would provide guarantee against arbitrary increase of price by the suppliers. The argument against this was that it would cost some billions of dollars to create this "excess" production capacity. And clearly from that point of view it should have been done rather than to get ourselves into the present pickle. But it wouldn't have helped of course at all in the long-term energy picture.

In the fall of 1973 came the Arab OPEC embargo followed by high prices, which have nothing to do with the embargo. The suppliers just see it in their interest to get more money for the goods which they have to sell. This brought into existence the Federal Energy Office and a lot of complicated regulations. Also initially long lines at gasoline stations. Unfortunately, my understanding is that we still have no policy in this country to deal with the energy problem, either short-term or long-term, except high price. High price is to my mind a tool and not an end, although for people who have oil to sell it may be an end.

What I want to do now is to look at some aspects of US national energy policy and show some of the hazards and perhaps indicate some means to improve what we're doing. The decision to restrict gas station operating hours in the fall of '73 rather than to ration gasoline didn't take into account the overall social costs involved. It did make consumers believe there was a problem. And there <u>is</u> a problem. There is a second problem of assurance of supply. There is a problem of long-range energy supply for this country. But this latter terrible problem I think isn't the one that people usually consider.

Let's look at one aspect of energy policy, at the regulations which set the price for old oil. This said that old oil (which is oil from fields which were in production before 1972) could be sold at not more than $5.25 a barrel, and the same regulations decontrolled new oil, which sells for about $11 a barrel. Furthermore the regulation stipulated that for every barrel of new oil produced, the producer was entitled to release a barrel from control and to sell it at the uncontrolled market price. That's history. The Ford Administration wanted to decontrol the price of old oil as well as new oil, and argued that there would thus be a greater incentive to produce more oil. I'm discussing this specific question in the context of setting policy and I'm talking about understanding the consequences of different candidate policies. It is easy to show without a computer model that the consequences of the Administration policy are very different from what was claimed, and therefore we ought to do this kind of modeling in a serious and objective way in order to evaluate candidate proposals and to select one which works. But it's clear that under the old regulations to provide a barrel of new oil and sell it at $11 a barrel (and as a consequence be able to release a barrel of old oil and sell it too at $11 instead of $5 a barrel) amounted to an incentive of $17 per marginal barrel of new oil. Under the decontrol proposal the marginal incentive would have been just $11. (Less the same costs in both cases, of course.) You produce a barrel of new oil, you sell it at the market price; that's it. Under any reasonable model of behavior of producers, decontrol would have produced windfall profits on old oil. I'm not against profits; I'm not even necessarily against windfall profits. But that wasn't the stated purpose of the policy change. The purpose was said to be to create a greater incentive to produce new oil, and it would not do that. It would have reduced the incentive to produce new oil.

The point may be moot now because of recent legislation which mandates an average price for new and old-oil. I don't think anybody knows what the incentive is as a consequence of this legislation. I raised this point in a September 1975 discussion paper for the Environmental Advisory Committee. [This is a committee under the Federal Advisory Committee Act and so our meetings are attended by the press, by industry, by anybody who wants to come to observe the Federal Energy Administration Environmental Advisory Committee. The audience can send us propaganda or nasty letters, as the case may be. So I feel no reluctance in sharing with you the contents of this paper.] I raised these questions as to the incentive in the Administration decontrol proposal and found pretty general agreement among economists in this country and even among FEA staff. So the question is how we get a policy or a policy proposal from FEA and from the Administration which does what it is supposed to do.

WHAT'S GOOD ABOUT HIGH PRICES?

Let's look ourselves at the purpose of high prices. There are, so far as I can see, three benefits that have been ascribed to high oil prices.

The first is that they would call forth alternative energy sources, for instance synthetic crude oil from coal, gasified coal, solar energy, and the like. The second is that they would provide capital formation to all the necessary investment to produce more oil or these alternative energy sources. And the third good is that they reduce demand.

But if you look at these three benefits, only the third is unambiguous. High prices really do reduce demand eventually; the short-term elasticity may be fairly small, but the long-term is I think pretty big. But this reduction of demand could be done equally well by a product tax, and so there is an alternative. Don't raise the price of oil to the producer necessarily; raise the price of oil to the consumer if you want to reduce demand. And the choice between these alternatives should have been made on an informed basis.

The second argument for capital formation is questionable in principle and in practice. In principle because that's not the way you get money to do something in this country. You have a prospect which looks good to investors, you sell stock to get equity capital, you borrow money, and you're in business. If only those who are already in the business can get capital to expand production or to go into alternative energy sources, that's in principle a bad thing and changes our entire system of operation in this country. In practice it doesn't work either, because Mobil in 1974 put almost a billion dollars of its money into the purchase of Montgomery Ward (MARCOR). If there's a big capital shortage in the oil companies, why didn't Mobil put that money into expanding production of oil or into alternative energy sources? So I say the second reason for high prices is at best questionable.

The first reason; namely, to call forth alternative sources of energy, would be available equally well under high product prices produced by a tax, but it is not available really under a decontrol system where the price of oil is set by an extralegal cartel--by OPEC. No US law that I know of keeps OPEC foreign countries from setting any price that they want. And there is no law or custom which keeps domestic producers from charging that same price. That's their option. They can't charge a higher price because Saudi Arabia has plenty more oil to sell and we wouldn't sell any here. They don't want to sell at a lower price. They can sell anything they can produce right now at the OPEC price, and that will be the case until potential domestic energy production could satisfy US needs. The problem with bringing forth alternative sources of energy like synthetic crude and coal gasification and solar energy is that there is no guarantee that energy in the future will sell at this high price, and in fact if the whole world goes over to solar energy (let's fantacize) at \$5 per barrel equivalent, then the oil producers are not going to sit there and not sell their oil at all; they will lower the price. They may lower the price enough to put these alternative sources out of business. This is a major reason why you don't see much investment in these alternative sources. So whether or not there is twenty years or forty years of oil left, oil producers and other investors have much more incentive at present to put their money into more oil and to exhaust the supply faster than into coal-based gas or synthetic crude. The incentives are not there under this system (even if the money is) to call forth alternative sources of energy.

There are other problems with our policy and regulations. These problems are ones which were predictable and have not yet necessarily been observed, but I feel that they are there and will come to light. I raised some of them in December 1973, claiming that this two-price system for oil without vastly larger civil and criminal penalties would lead to various peculiar operations. For instance a tanker could load with old oil in Texas at $5 a barrel and it could arrive in New York, perhaps stopping at some intermediate port, with the manifest reading: Libyan oil at $11 a barrel. Instead of earning 50 cents a barrel for transporting the oil, a 50,000 ton tanker thereby earning $250,000 for the trip, could make almost $6 a barrel and earn $3 million for the trip. Has this happened? I don't know. I don't think businessmen in general are dishonest, but the opportunities here are so high that honest businessmen will be bought out by fronts or by would-be criminals in view of the small fines that could be imposed for such activities.

HOW DETAILED THE MODEL?:

So in my opinion a model has to be sufficiently detailed (disaggregated) to make provision for such activities. Need we include all the 200 million people in this country in a fast-time computer model? We certainly don't have to include every barrel of oil, because the oil can be aggregated into tanker loads at least. People cannot be, because you don't know which person is going to be the entrepreneur who has the idea which is going to put the system on its ear. It is the possibility of exercise of free will which is the problem of modeling people.

Let's first look at the producing sector and ask to what extent that could be aggregated in a model which would be suitable for testing candidate national energy policies. We would have to be careful not to mask by aggregation qualitatively or quantitatively important effects. If I have two balloons and I say they are equivalent to one balloon, without making sure that they are equivalent by stabilizing the system, I would get a very wrong answer. And if I put all of the production together into one producing sector, I think I will get a very wrong answer. So I'll have to consider separate producing sectors. I'll have to consider first major domestic oil producers with large oil supplies abroad. Their behavior is likely to be different from a second group of major domestic oil producers with little foreign oil but maybe with large coal holdings. And a third group: coal producers. And a fourth group: potential coal producers--organizations with money, credit, and management talent. It may be that this last group is most important to the goals of Project Independence. But I don't know because such detailed modeling has not been done.

It's rather a big thing to specify such a model and to write such a model. The Federal Energy Administration has a model called the Project Independence Evaluation System (abbreviated "PIES"), formed by bringing together sub-models from all over the government and academic institutions. The results of the model are available and have been criticized by a number of groups. My September paper was a partial criticism for unwarranted aggregation and for inconsistent assumptions. For instance, arguing that the decontrol of old-oil prices will have very little effect on overall energy costs (because natural gas is price controlled and the coal is mostly under long-term contracts) is inconsistent because it was the policy of the Administration at the same time that it was pushing for decontrol of oil prices to push decontrol of gas prices. And so you have to assume some coherence. If you succeed in one decontrol, you are likely to succeed in the other. Inconsistency in regard to the coal long-term contract, because for instance it was just at that time that Westinghouse told utility customers that had firm contracts to buy uranium from Westinghouse at about $10 per pound that it would no longer supply uranium at this price because the overall exposure to Westinghouse would be about $2.5 billion if it had to fulfill those contracts over the next several decades. For the FEA, knowing this, to rely on the sanctity of the long-term coal contracts seem to me to be disingenuous. Maybe ingenuous.

IMPORTANT SECTORS IGNORED:

There are some other sectors of society which have to be included in any model. These sectors exist; they're important because they get paid lots of money; they're visible in the newspapers, and yet they don't show up in the model. For instance, lawyers. If it's important to pay lawyers, it's important to model lawyers. What is the influence of the legal profession and the courts in energy policy? Criminals--some grain inspectors in Baton Rouge have been indicted recently and charged with taking $67,000 in bribes, which reduces the value of shipments by many millions of dollars because of the loading of imperfect grain. Such actions have a big effect on the system, like termites boring invisibly at the foundations. And the model ought to include them.

Now it's very hard for the systems analyst to be as clever in all the same ways as lawyers on the one hand or criminals on the other. And so you can't rely exclusively on the model designer. Economists don't really have the necessary bent of mind. It seems to me what one has to do is to have a model such that there is room for human initiative. There could be some terminals operated by people who are motivated. Perhaps college students, undergraduates, are the best ones to be paid (in fact you don't probably have to pay, they'll pay you) to act the role of an imaginative criminal. Law students can be asked to show how by delay they can postpone the imposition of penalties until they make no difference. It is very important for the people section of models to be disaggregated to the point where these important sectors can be included.

Now how about people in general? Maybe some people are going to not pay their income tax and other people are going to have great ideas for founding companies. Do we have to model every person in this country, N of them where N is 200+ million with a Monte Carlo probability of P (where P is very small so that PN is a few) in order that we be able to catch these few important events? Of course not! So long as the interest is in a few rare events and not in a coincidence of rare talents, a different sector can be included in the model, (a criminal sector or an ingenious sector or the like) where the number of individuals N is reduced by a factor F and the Monte Carlo probability P increased by a factor F, so that the event rate of the extraordinary events remains the same.

CONCLUDING REMARKS:

I've touched on the problems of aggregation in modeling electrical power systems; on balloons and gas storage; on time-of-day charging, and the details which have to be considered before you can be sure what that's going to do; on national energy policy. The same detailed modeling is probably necessary to the setting of tax policy, of educational policy. Having listened to simulation proponents and to automatic control proponents and to those who think that if everybody is just better educated we will automatically arrive at a better country, I think that among our realistic choices it would be very useful to have more simulation, more understanding of what happens, more testing of candidate proposals and policies against the discipline of computer simulation. These simulations ought to be a lot less aggregated than they are and should make provision for real-time interaction with ingenious people who don't know anything about computers but who have good ideas how to manipulate the system eventually to their own advantage. Finally, a simulation should be presumed defective which does not model the behavior of a sector of society which figures importantly in the real world.

IBM THOMAS J. WATSON RESEARCH CENTER, POST OFFICE BOX 218, YORKTOWN HEIGHTS, NEW YORK 10598.

Proceedings of Symposia in Applied Mathematics
Volume 21
1977

PROJECT INDEPENDENCE EVALUATION SYSTEM: STRUCTURE AND ALGORITHMS

William W. Hogan*

Presented at:
American Mathematical Society Short Course
on Mathematical Aspects of Production
and Distribution of Energy
San Antonio, Texas
January 20-21, 1976

Abstract

The development of mid-term energy forecasts for the United States is accomplished under the title of the Project Independence studies of the Federal Energy Administration. The analytical framework for Project Independence can be viewed as a large model drawn from the techniques of many disciplines. The structure of the model is overviewed with the emphasis on the optimization and fixed-point formulations and algorithms. Computational experience is indicated.

---

* Formerly of the Federal Energy Administration;
Currently at Stanford University, Department
of Engineering-Economic Systems and
Department of Operations Research

I. Introduction

From its inception, the Federal Energy Administration (FEA) has dealt with a variety of national energy policy issues. Typically, the evaluation of policy alternatives involves the assessment of the near-term evolution of the energy system. A comprehensive summary of these studies is found in the now annual series of publications initiated with the Project Independence Report [10] and the National Energy Outlook [8]. The FEA employs a large-scale modeling system to organize the data and prepare the conditional forecasts. This modeling system incorporates a variety of methodologies and computational procedures. The purpose of the present paper is to overview the mathematical statement of the model, the interpretation as a fixed-point problem, and the special algorithms exploiting the problem structure. The computational experience is summarized.

II. Policy Environment

The environment for the initial evolution of energy policy models at FEA was characterized by (i) a rapid increase in concern for the details of the system interactions, and (ii) the active participation of interests groups with diverse perspectives.

The disruptions of the world oil market in 1973-1974 found the United States government without a comprehensive energy policy. Further, the existing mechanisms for the evaluation of the component policy issues were not adequate. The traditional experts were few in number and were without the clairvoyance that would have prepared them to analyze the unexpected dislocations caused by the oil embargo or the rapid increase in energy prices. Previously, the general discussion of national trends constituted the primary Federal interest in energy forecasting. Abruptly, the requirement arose for the description of the interaction of multiple energy product supply and demand over time with a variety of geographical classifications. Extensive data were available but there was no procedure for organizing the information or testing the implications of alternative future scenarios. These circumstances provided the opportunity for the fruitful application of formal modeling. If the models could be constructed with the speed consistent with the perceived urgency of the problem, they would be highly valued and used extensively. The systematic organization of the questions and the data would provide structure for the policy debate. Computer implementation of a sound framework would permit some approximation of the detail being requested. The constructive model solution would provide support for the response to the difficult challenges to credibility confronting every forecast or policy statement. With these expected benefits, a major effort was initiated to develop the necessary models and the attendant analytical capability.

The structure of the models was determined by the nature of the problem and the character of the potential consumers of the modeling results. The energy sector is pervasive in its impacts on our economy. The production, distribution, and consumption of energy products involve the activities of many industries. When examined in detail, the energy system can be quite complex. It is not possible to identify a single element of the system which characterizes all the important policy variables. Some components may receive highlighted attention, e.g. imports, but an essential feature of any energy model is the description of a multi-attribute system. The interests of the potential consumers were similarly diverse, and there was little unanimity in defining desirable futures that were also feasible. Without this definition of a criterion, a normative modeling approach was not sought nor would one have been used. The demand was for a descriptive model that calculates the logical implications of a consistent set of assumptions. This scenario approach opened the utilization of the model to a spectrum of users. The value of the model is judged by the empirical content, the ability to reduce the number of input assumptions, and the ability to demonstrate the plausibility of any resulting forecast. This descriptive perspective was adopted in the construction of many models within the FEA. In the remainder of this paper, we concentrate on the most extensive of these models, the Project Independence Evaluation System (PIES).

III. The PIES Formulation

After turning its attention from the immediate problems of the oil embargo, the FEA began an extensive and systematic investigation of the prospects for energy self-sufficiency under the title of "Project Independence." This effort expanded quickly into the evaluation of an array of policy issues confronting the United States in the next decade. The province of the formal model evolved into the regional characterization of the production, transportation, conversion, and utilization of multiple energy products given explicit assumptions about the economy, import prices, and national policies. The detailed problem characteristics of PIES are described elsewhere [4],[8],[10]. The system is actually a highly modular collection of specialized component models. For the present, it sufficies to formulate the problem at the level of the PIES integrating framework using the representation of a network as shown in Figure 1.

The organizing principle of the model is found in traditional microeconomics. The potential level of production of any energy product is assumed to be sensitivie to the price paid for that product, with higher prices inducing greater output. Conversely, the demand for any energy product is assumed to be inversely related to the price paid. An equilibrium is obtained whenever a price is determined such that the quantity demanded and the quantity supplied are the same. In Figure 1, this idea is extended to the consideration of numerous products, geographically dispersed, and connected by a system of transportation and conversion activities. At each supply region, for each product, a separate model is employed to characterize the price quantity relationship for the output of that particular product. Once produced, the energy product is moved through a transportation and distribution system where, in competition with other energy products, it may be converted into some alternate energy form, e.g., coal used to make electricity. After such conversion or refining, the products are transported to the demand regions. In the demand regions, a separate model is employed to characterize the price quantity relationship determining the demand for the energy products. A key feature of this component is that the demands for any product are determined from the prices of all the competing products. In the jargon of economics, this is the explicit recognition of the cross price elasticities of energy demand.

Implicitly, the characterization of a solution incorporates the cross price effects of the supply activities. The detail of the individual supply curves and the distribution network combine to provide a representation of the activities necessary to supply each product in each demand region. If the quantity to be delivered to the demand regions is known, the competitive selection of supply alternatives is equivalent to that required to deliver these quantitites to the demand regions at minimum total cost. This equivalence between competitive solutions and minimum cost supply is a familiar result in economics and a central element of the model formulation. Furthermore, the change in the total cost required to deliver an additional unit of a given energy product can be interpreted as the supply price of that product. The minimum cost solution and its derivatives, therefore, provide a complete characterization of the implicit supply curves for energy products delivered to the demand ergions. A detailed example of the implemen-

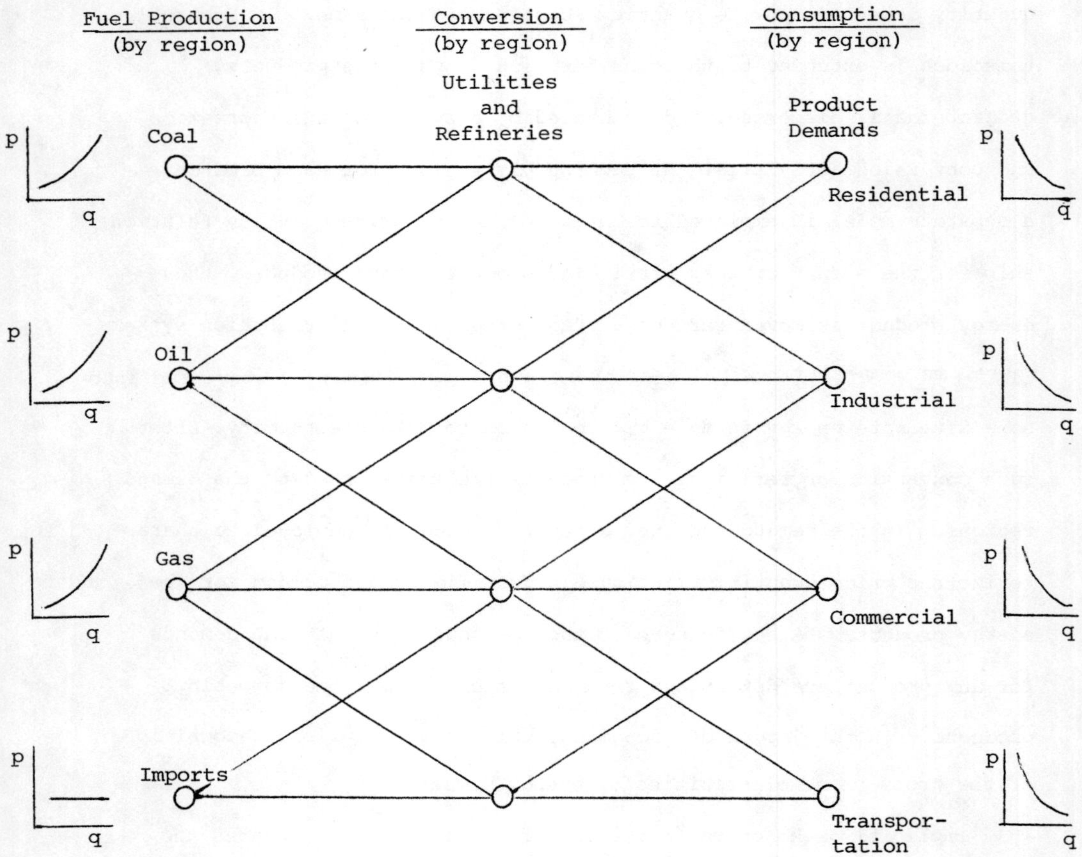

Figure 1

The Energy System

tation of this idea can be found in [4].

If the quantities of energy products delivered to the demand regions can be determined such that the associated supply prices and demand prices are equal, we have obtained the generalization of the single commodity equilibrium solution. Such a solution is interpreted as our forecast of the future state of the energy system. It provides an extensive representation of the details of the flow and utilization of energy products. For example, with the price of imported oil set at a fixed level, the model can be used to trace out the substitutions of other energy products and the resulting demand for imports. If the quantity of imports is fixed, the model determines the equilibrium price or tax that is needed to reduce demand to the desired level. By altering the input assumptions for the generation of the supply and demand curves, the implications of energy policies can be traced throughout the system in a manner which preserves a sound theoretical framework while manipulating a large volume of data. Hence, by careful description of this network, the essential forecasts can be obtained for the policy evaluation input.

This formulation of the energy system in terms of the equilibrium solution of a network of supply and demand activities is a natural approach to the problem. Similar approaches for the full system or major components can be found in [2],[7]. The assembly of the network involves a large data collection effort as discussed in [8]. Although the definition of the solution of the resulting network is straightforward, the search for a solution requires some special methods. Before turning to the description of these methods, we convert the general description of the model of Figure 1 into a formal mathematical statement.

The description of the supply system as a network is convenient at the level of detail in Figure 1, but the system does not satisfy the formal definition of a network. Many commodities flow through the system and there is substantial interaction among these commodities. In particular, joint inputs and outputs are required for the conversion processes and the possible combinations of these joint products are restricted to some convex set. These complexities preclude the computational advantages of simple networks, but they can be handled within the framework of convex optimization. In theory, the supply problem of providing the desired combinations of energy products in the demand regions at minimum cost is always a convex program. In practice, this problem lends itself to a natural and acceptable linear programming approximation. We represent this as

$$\begin{aligned} &\min_{x \in X} \quad cx \\ &\text{st. } Ax = q \end{aligned} \qquad (1)$$

where

x : the vector of supply activities,

c : the vector of costs of supply activities,

X : the feasible set of supply activities,

A : the matrix of accounting relationships mapping the supply activities into energy product demands,

q : the vector of energy product demands.

In this application, the vector x has dimensions in excess of 4000. A particular element of x may be the level of operation of some refinery, the next increment on a supply curve, a transportation activity, etc. The feasible set X is characterized by over 1000 linear equations defining the physical properties of the supply system. The matrix A accumulates the shipments into the demand regions to match with the vector of demands for energy products in each region q with dimension 100. If q is known, the competitive supply is determined by the solution of Eq.(1). From standard linear programming theory, there exist a set of dual variables π associated with the demand constraints in Eq.(1). These dual variables are interpreted as the supply prices.

The demand side of Figure 1 is represented by a vector valued function which maps the prices p of the demand products into the associated quantities and is denoted as

$$q = q(p) \quad . \tag{2}$$

If $\pi(q)$ denotes the optimal dual solution of Eq.(1) given demand quantities q, then the equilibrium solution of the energy system problem is a set of dual variables, demand prices and quantities such that

$$\begin{aligned} \pi^* &= \pi(q^*) \\ q^* &= q(p^*) \\ p^* &= \pi^* \quad . \end{aligned} \tag{3}$$

The next section pursues the solution of Eq.(3) in the presence of certain simplifying assumptions.

IV. Formulation as an Optimization Problem

The demand functions defined in Eq.(2) are derived from an econometric model and possess economic properties which can be exploited in the pursuit of a solution for Eq.(3). The "direct demand" function of Eq.(2) is invertible. Hence, we can speak equally of a set of demand prices determined by the quantities. This "indirect demand" function is

$$p = p(q) \quad . \tag{4}$$

With this definition, the last condition in Eq.(3) becomes

$$p^* = p(q^*) \ . \tag{5}$$

These direct and indirect demand functions are assumed to possess general properties of traditional economic goods. In the present case, this implies that the derivative of p(q) is a negative definite matrix for all q. This is a weak economic assumption and an observed property of the PIES demand functions over the relevant regions. With these conditions on the demand function, and motivated by the implicit characterization of the supply functions in Eq.(1), it is natural to interpret Eq.(3) as the solution to a related convex optimization problem. This convenient formulation can be obtained with one additional assumption about Eq.(4).

Suppose that p(q) is an integrable function in the sense that the line integral of the inner product $p(q)\cdot dq$ between any two points is independent of the path of integration. Then there exists a function g(q) such that

$$g(q) = \int_0^q p(\bar{q})\cdot d\bar{q} \ , \tag{6}$$

with

$$\nabla g(q) = p(q) \ , \tag{7}$$

and,

$$\nabla^2 g(q) = \nabla p(q) \quad . \tag{8}$$

With these observations, we formulate the problem

$$\begin{aligned} &\underset{\substack{x \varepsilon X \\ q}}{\text{Min}} \quad cx - g(q) \\ &\qquad \text{s.t.} \quad Ax - q = 0 \ . \end{aligned} \tag{9}$$

Now if q^* is an optimal solution for Eq.(9), it follows directly that $(\pi(q^*), q^*, p(q^*))$ is a solution of Eq.(3). Conversely, any q^* obtained as a solution of Eq.(3) is optimal in Eq.(9). The formulation in Eq.(9) has substantive and computational appeal. The negative of the objective function is the difference between the cost of providing a given level of supply and the consumer surplus measure of the value of that supply. This has a natural economic interpretation as a net social welfare function. The economy can be viewed as optimizing the value of this welfare measure [11].

The negative definiteness of $\nabla p(q)$ implies that $g(q)$ is a concave function and, therefore, Eq.(9) is a convex optimization problem. This permits the application of known and well-behaved algorithms. In many applications, $p(q)$ is a linear function, producing a quadratic program in Eq.(9), e.g., in [7]. More often, adequate piecewise linear approximations for $g(q)$ are available. This converts Eq.(9) into a tractable linear program. In this context, the implicit nature of the supply functions in Eq.(1) is no handicap, and the direct solution of Eq.(3) can be obtained through standard methods.

It is not surprising, therefore, that this optimization formulation has been pursued frequently in related problems. It simplifies the interpretation of the fundamental properties of the solution and exploits powerful computational procedures. Unfortunately, the assumptions supporting this representation of the problem are not satisfied by PIES. However, we develop below an alternative ad hoc solution procedure which is motivated by Eq.(9) and systematically exploits its most attractive computational features.

V. Reformulation as a Fixed Point Problem

The difficulty in the series of assumptions needed to develop Eq.(9) is found in the integrability of $p(q)$. It is well known that the second derivative of a continuously differentiable function of many variables is symmetric. From Eq.(8), therefore, it follows that $\nabla p(q)$ must be a symmetric matrix. However, the demand functions generated within PIES are not symmetric. Therefore, the integral in Eq.(6) is not well defined and the generalized measure of consumer surplus $g(q)$ does not exist. This unintuitive and seemingly technical point is important for theoretical and computational reasons. It is prevalent in the literature on demand functions with more than one variable. The exposition of the theoretical implications is not pursued here, but the interested reader is referred to [6], or [9]. The computational implications are clear. The convenient equivalence between Eq.(3) and Eq.(9) is no longer available. Although the problem in Eq.(3) continues to be well defined as the proper equilibrium solution of the energy system network, some alternative procedure must be designed to obtain a solution.

The perspective adopted here is to reformulate Eq.(3) as a fixed point problem. This leads to an iterative procedure for obtaining the intended fixed point. Recalling Eq.(1), we define the extremal value function, $v(q)$, as

$$v(q) = \min_{x \in X} \; cx \quad \text{s.t. } Ax = q \; . \tag{10}$$

From convex duality theory [3], the optimal dual solutions or the energy supply prices are the subgradients of $v(q)$. We define the point-to-set map of supply prices for given quantities as

$$M(q) = \left\{ \pi \,\middle|\, v(q') \geq v(q) + \pi(q'-q) \right\} \; . \tag{11}$$

Using $M(q)$, it is possible to define a sequence of calculations starting with the demand prices p and terminating with the supply prices π . The resulting point-to-set map of supply prices is denoted as $S(p)$:

$$S(p) = M(q(p)) \quad . \tag{12}$$

In terms of $S(p)$, the fixed point formulation of Eq.(3) is the determination of a p^* such that

$$p^* \in S(p^*) \quad . \tag{13}$$

With this view of the equilibrium problem, the natural approach to a solution is an algorithm generating a sequence $\{p^t\}$ where

$$p^{t+1} \in S(p^t) \quad . \tag{14}$$

This is the essential idea of the procedure below, but some modifications to Eq.(14) are necessary. In its current form, it is a "cobweb" approach for which there are many examples of divergence. In the present context, the difficulties are compounded by the fact that a linear programming algorithm is employed to determine an element of $S(p)$. Typically, this will produce an extreme point element of $S(p)$, a property which may exclude the solutions of Eq.(13). This situation is illustrated in Figure 2 for the determination of the equilibrium price and quantity for a single commodity. Two examples of $S(\cdot)$ are included. At the price p' the corresponding set of supply prices obtained from the approximate supply curve is the singleton $S(p')$ but it is not an equilibrium. However, for the equilibrium price p^*, $S(p^*)$ is an interval and the extreme points are not equilibrium solutions. This difficulty is overcome by including some information about the demand function in a sequence of problems successively approximating Eq.(13). Each problem in this sequence of approximations is of the form of Eq.(9) for an appropriate simplification of the demand function. Hence, the resulting procedure combines the rigorous theory of the formulation in Eq.(13) with the attractive computational power of Eq.(9).

VI. Approximation Procedure

The difficulty in Eq.(9) centers on the existence of the function $g(q)$. This is determined by the fact that $\nabla p(q)$ is not a symmetric

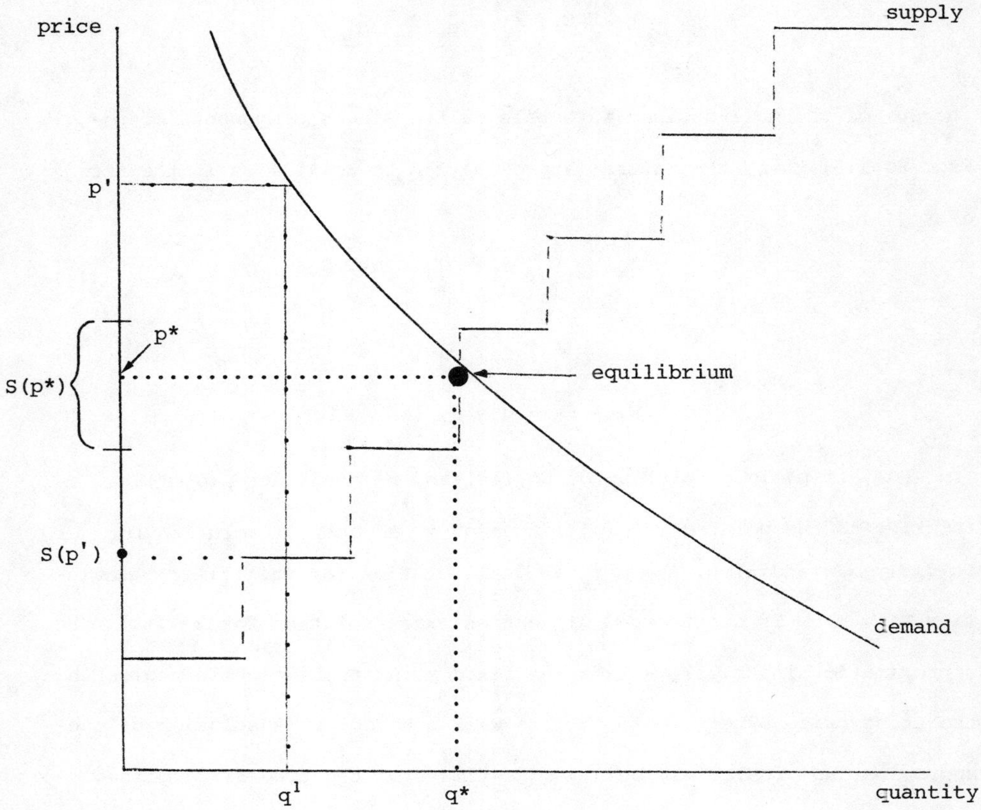

Figure 2

Equilibrium As A Fixed Point

matrix. By employing a demand function which approximates p(q) but has symmetric derivatives, a problem of the form of Eq.(9) can be recovered and it should be a close approximation of Eq.(13). This is the essential ad hoc step in the approximation method.

The approximation defines a symmetric indirect demand function $\hat{p}(q,\hat{q})$ relative to the point $\hat{q}$ via the n components of p as

$$\hat{p}_i(q,\hat{q}) = p_i(\hat{q}_1, \hat{q}_2 \ldots, q_i, \ldots \hat{q}_n) \text{ , } i = 1, 2, \ldots n. \quad (15)$$

The indirect demand function $\hat{p}(\cdot,\cdot)$ intersects $p(\cdot)$ at $\hat{q}$, i.e. $(p(\hat{q}) = \hat{p}(\hat{q},\hat{q}))$ and possesses the same own price effects as $p(\cdot)$. The cross price effects are zero, however, and $\nabla\hat{p}(\cdot,\hat{q})$ is therefore trivally symmetric with all negative diagonal elements. For this demand function, Eq.(6) becomes

$$\hat{g}(q) = \sum_{i=1}^{n} \int_0^{q_i} \hat{p}_i(\bar{q}_i, \hat{q})\, d\bar{q}_i \,, \tag{16}$$

the sum of the well-defined integrals of the single components of $\hat{p}(\cdot, \hat{q})$. From Eq.(16) then, the natural approximation to Eq.(13) is in the form of Eq.(9) as

$$\begin{aligned} &\underset{\substack{x \in X \\ q}}{\text{Min}} \quad cx - \sum_{i=1}^{n} \int_0^{q_i} \hat{p}_i(\bar{q}_i, \hat{q}) \cdot d\bar{q}_i \\ &\quad \text{s.t.} \quad Ax - q = 0 \;. \end{aligned} \tag{17}$$

Now, if p^* is a solution of Eq.(13) and we construct $\hat{p}(q,q^*)$ relative to the point $q^* = q(p^*)$, then q^* is an optimal solution for Eq.(17). In addition, the optimal dual solution for Eq.(17) π^* satisfies $\pi^* = p^*$. If constructed without an exact solution for Eq.(13), the approximation in Eq.(17) solves the fixed point problem while neglecting the cross price effects. If $p(\cdot)$ is well behaved, this solution of the approximating problem may be a good estimate of the true fixed point.

This conjecture leads to the algorithm:

Step 1: Choose a vector of demand prices, p^1. Let $t=1$

Step 2: Calculate $q^t = q(p^t)$. Construct $\hat{p}(q,q^t)$ and Eq.(17) relative to the point q^t. Solve Eq.(17) for the optimal (x^t, π^t).

Step 3: If $\pi^t = p^t$, go to Step 4. Else, let $p^{t+1} = \theta p^t + (1-\theta)\pi^t$ for some $0<\theta<1$. Go to Step 2.

Step 4: Terminate with the equilibrium supply pattern x^t, consumption q^t, and market prices p^t.

As implemented, the repeated solution of Eq.(17) can be accomplished exploiting the reoptimization capabilities of linear programming codes. It is not necessary to obtain the exact integrals in Eq.(17), only the changes in the integrals relative to the point q^t are needed to preserve the properties of the approximating problem. With the application of a grid partition and the piecewise linear representation of the integrals, Eq.(17) becomes a linear program and each iteration is captured by changing a few elements in the objective function and the constraint limits. The details of this further simplification are found in [4].

VII. Computational Experience

The theoretical convergence properties of this approximation procedure in the presence of general demand functions have not been established. The well-behaved properties of related processes for general equilibrium models are discussed at length in [1]. Unfortunately, the PIES model does not satisfy the gross substitutability or homogeneity assumptions needed for those results. Quadratic optimization and complementarity approaches are indicated in [13] which contains the most comprehensive discussion of such formulations and some relevant computational experience.

The fixed point formulation in Eq.(13) requires the solution of a complicated problem of dimension 100. Since the evaluation of the function involves the solution of the underlying supply optimization problem, the expense of solving the model with general fixed point approaches [12] would be prohibitive. In practice, the ad hoc procedure has worked with remarkable speed and robustness. A wide variety of problems are solved on a routine basis at the FEA. Typically, 6 - 10 iterations of the algorithm are adequate in obtaining the desired fixed point within the limits of the linear approximations. Because of the efficient reoptimization of Eq.(17), the time required to complete the full solution is no more than twice that required to obtain the first linear program solution from a cold start. The system has never failed to converge and, because the termination test is a constructive test of an equilibrium, the method is fail-safe. This remarkable success is not fully understood. It is probably two or more orders of magnitude faster than competitive general fixed point approaches, and should be considered seriously by others faced with similar problems despite the current lack of a full theoretical exposition.

The robustness of the procedure has stimulated a variety of computational experiments. Most prominent is the inclusion of average cost pricing. In many cases, the equilibrium in the energy system is between the demand price and the average cost of delivery, not the marginal cost implied by π. This average cost can be determined as a function of the deliveries and their costs $f(c,x^t)$. If we substitute $f(c,x^t)$ for π^t in Step 3 of the algorithm, the procedure is well defined and the average cost equilibrium is obtained. This important substantive change severely complicates the mathematics of the problem but has little effect on the observed convergence properties of the algorithm. In the few cases where convergence of this more sophisticated method has failed, ad hoc adjustments were available to recover the solution. Again, the fail-safe nature of the termination criterion prevents the most serious difficulty and the computational advantages are preserved.

The success of this iterative algorithm was balanced by the difficulty of organizing and manipulating the data for the 1000 x 4000 linear program constituting Eq. (17). This is an important issue in the discussion of the computational properties of the full system, but it is beyond the scope of the present discussion. The interested reader is referred to [4], and [5].

VIII. Summary

The PIES model employs an integrating framework grounded in the basic principles of economics. It has applicability to many other policy problems. This paper has presented an overview of the mathematical structure of the model and the special algorithms designed for its solution. These algorithms exploit optimization and fixed point interpretations of the model. They have been extremely successful in dealing with a large and complex problem. The computational experience recommends the approach for broader application and deeper theoretical study.

REFERENCES

1. Arrow, K.J., H.D. Block and L. Hurwicz, "On the Stability of Competitive Equilibrium - II," Econometrica, vol.27, 1959, pp.82-109.

2. Energy Hydrocarbons in the United states to 1985, the PACE Company Consultants and Engineers, Inc., Houston, Texas, January 1974.

3. Geoffrion, A.M., "Duality in Nonlinear Programming: A Simplified Applications Oriented Development," SIAM Review, vol.13, 1971.

4. Hogan, W.W., "Energy Policy Models for Project Independence," Computers and Operations Research, vol.2, Pergamon Press, 1975.

5. Hogan, W.W., "Matrix Generation, Report Writing, and the Scale of Future Applications," paper presented at the Miami ORSA/TIMS Meeting, November 1976.

6. Hurwicz, L., "On the Problem of the Integrability of Demand Functions," in Preferences, Utility, and Demand, ed. by John S. Chipman, Harcourt Brace Jovanovich, New York, 1971.

7. Kennedy, M., "An Economic Model of the World Oil Market," The Bell Journal of Economics and Management Science, vol.5, no.2, Autumn 1974.

8. National Energy Outlook, Federal Energy Administration, Washington, D.C., February 1976.

9. Pressman, I., "A Mathematical Formulation of the Peak Load Pricing Problem," The Bell Journal of Economics and Management Science, vol.1, no.2, Autumn 1970.

10. Project Independence Report, Federal Energy Administration, Washington, D.C., GPO No. 4118-00029, November 1974.

11. Samuelson, P.A., Foundations of Economic Analysis, Atheneum, New York, 1965.

12. Scarf, H. and T. Hansen, The Computation of Economic Equilibria, Yale University Press, New Haven, Connecticut, 1973.

13. Takayama, T. and G.G. Judge, Spatial and Temporal Price and Allocation Models, North-Holland, Amsterdam, 1971.